War, Technology, and Experience
aboard the USS *Monitor*

War, Technology, and Experience aboard the USS *Monitor*

David A. Mindell

The Johns Hopkins University Press
Baltimore and London

The Johns Hopkins University Press
2715 North Charles Street
Baltimore, Maryland 21218-4363
www.press.jhu.edu

Library of Congress Cataloging-in-Publication Data
will be found at the end of this book.

A catalog record for this book is available from the British Library.

ISBN 0-8018-6249-3
ISBN 0-8018-6250-7 (pbk.)

Frontispiece photograph: The low freeboard of the monitors made them part submarines, as their machinery and crew spaces were completely below water while at
sea; only the turret *(left)* and the pilothouse *(not shown)* protruded above. This is the
USS *Terror* in a photograph from 1898, but the situation was much the same with the
original *Monitor* in 1862. Courtesy of the Naval Historical Center, Washington, D.C.

To Merrit Roe Smith and Leo Marx,
my best teachers, sharpest critics,
and sources of inspiration

Contents

Illustrations

Preface and Acknowledgments

Usually we read about the *Monitor* as the story of a heroic inventor and a revolutionary new warship. I expand that story to include patrons, contractors, constructors, rivals, users, public imagery, and literary expressions. Adding these dimensions to the well-known history changes our understanding of the event and its significance in particular and also serves as an introduction to ideas in the history of technology more generally. The USS *Monitor* story provides a lens through which to see issues of technology and society, military technology, and the human implications of new machinery. These questions place the ship within the larger history of the mechanization of warfare, a history not usually studied along axes of experience and representation.

I do not repeat the enormous literature on the *Monitor* but make use of it wherever I can, looking to reliable analyses for background and then calling on primary sources to make the major historical points. *Monitor* enthusiasts will note the absence from this text of a number of the standard anecdotes; the battle itself is retold in only a few paragraphs. Furthermore, this book has little to say about the CSS *Virginia,* though I do not suggest that the *Monitor* was the more important of the two vessels, nor even that it was more advanced. In one sense, a similar book might be written about the Confederate ironclad or, for that matter, about other ironclad ships and submarines built during the war for river and ocean service, addressing different but related experiences aboard those vessels. Still, because of its distinctive appearance (alternately threatening and comical), its connection to northern industrial infrastructure, its submarine qualities, and its status as a public symbol, the *Monitor* evoked the most discussion about the nature of modern warfare. One significant factor sets the *Monitor* apart from all other Civil War ironclads: the contrast between expectation and experience. No other contemporary warship combined such elevated celebration with such ambiguous results.

This project began six years ago as a paper for a seminar on early American technology taught by Merritt Roe Smith and Pauline Maier at MIT. I originally aimed to tease out the *Monitor*'s "real" place in the history of technology in contrast to its public image, but it soon became clear not only that that project had already been adequately done, but also that it was ultimately not consistent with our current understanding of technology as a social and historical enterprise. Two other events at the time also informed the work: the Persian Gulf War, in 1991, during which remote and unmanned technologies came increasingly into public view as determinants of a battle's outcome, and my own experiences in the world of oceanography, diving in deep-sea submersibles and submarines and living aboard cramped, poorly ventilated ships. The project thus grew into a second year paper as part of my graduate work in the history of technology at MIT. During a particularly intense semester, Deborah Fitzgerald, Leo Marx, Merritt Roe Smith, Sherry Turkle, and Rosalind Williams all contributed greatly to working through a complicated set of ideas. It ended up being published in *Technology and Culture* in April 1995, "The Clangor of that Blacksmith's Fray: Technology, War, and Experience aboard the USS *Monitor*." I wish to thank Robert C. Post for his generous help in bringing that paper to a publishable state and Barton C. Hacker (an anonymous reader who revealed himself to me) for helping to shape the book project in its early stages. In 1998 the paper won the Usher Prize from the Society for the History of Technology for the best paper published under the society's auspices in the previous three years.

The paper in *Technology and Culture* was but a small window into the rich material that surrounded the *Monitor*. In 1996, when I finished my dissertation (on an entirely different subject, the history of feedback control and computing), I decided to put that project aside and return to the *Monitor*. Bob Brugger of Johns Hopkins University Press was receptive to a proposal for a book that would go deeper than the paper and bring the ideas to a wider audience; he has shepherded the book through the laborious editing and review process. A number of colleagues, friends, and relatives patiently read the manuscript at various stages of completion, including Edward Eigen, Brendan Foley, Rebecca Herzig, Victor McElheny, David McGee, Phyllis Mindell, Marvin Mindell, Jessica Riskin, William R. Still Jr., and Tim Wolters. All have provided helpful commentary and discussion. The Mariner's Museum Library in Newport News, Virginia, kindly made the George Geer letters available to me. William R. Still, Jr., generously sent me copies of some documents from his personal collection.

War, Technology, and Experience
aboard the USS *Monitor*

Introduction

A Strange Sort
of Warfare

Like famished relatives after a funeral, when the firing stopped the crew of the USS *Monitor* sat down to eat. Their fight with the Confederate CSS *Virginia* had lasted nearly four hours on the clear Sunday morning of March 9, 1862. The *Monitor* survived the encounter relatively intact, and immediately the ship's steward went to work preparing Sunday dinner for the exhausted men. Visitors to the vessel that afternoon were astonished, wrote one officer, to find "a merry party around the table enjoying some good beef steak, green peas, &c."[1]

Soon, the assistant secretary of the navy, Gustavus Fox, came aboard and commended the crew: "Well, gentlemen, you don't look as though you were just through one of the greatest naval conflicts on record."[2] This sentence began a chain of pronouncements that continues to the present day. The *Monitor* and its crew leaped instantly into the national limelight, receiving the attention and praise of a troubled and soon-to-be-desperate nation. Within the week, Secretary of the Navy Gideon Welles predicted that the *Monitor* "must effect a radical change in naval warfare."[3] The press, the public, and the Union leadership hailed the *Monitor*'s performance not only as a military victory but also as a victory for new machinery, spelling the end of the "wooden walls" of the traditional navies of the world and the rise of superior steam-powered, armored fleets.

Americans celebrated both the technological change and its political implications, for they equated the older era of wooden warships with European (primarily British) dominance of the seas. The age of ironclads, it seemed, would belong to the United States, the land of invention, and so too would the international power and prestige that seemed to accompany a world-class navy. Congressional commendations, newspaper reports, and traditional histories echoed the celebrations.[4] In 1893, for example, the McCormick Harvester Company published a promotional poster that depicted the battle between the *Monitor* and the *Virginia* and tied its own machinery to that of the warships. "The fight settled the fate of the 'Wooden Walls' of the World and taught all nations that the War Ship of the future must be—like the McCormick Harvester—a Machine of Steel." These energetic, lasting responses made Hampton Roads a familiar icon of popular Civil War culture. Even today, the battle between the *Monitor* and the *Virginia* remains the most recognizable image of technology in the American Civil War. Some saw darker portents, as Henry Adams, writing from London, prophetically observed: "I tell you these are great times. Man has mounted science, and is now run away with. I firmly believe that science will be the master of man. The engines he has invented will be beyond his strength to control. Some day science may have the existence of mankind in its power, and the human race commit suicide, by blowing up the world."[5]

Indeed, Fox's commendation, which set off this chain of events, also hinted at an irony that shaped the crew's experience. "You don't look as though you were just through one of the greatest naval conflicts on record." What ought men look like after such a conflict? Fox never defined what he had in mind. It probably included some combination of torn uniforms stained with blood, hollow faces stunned by shellfire, and broken bodies black with powder. The *Monitor* crew had none of these defining marks (although the gun crew was covered with powder). They had fought the enemy while protected behind eleven inches of armor plate. Most had spent the entire encounter deep in the ship, below the surface of the water. None except the captain could even see the enemy. They did not look like naval heroes. For these men, the new machine brought a new experience of war.

The battle raised a profound question about the industrial world: in a war of machines, what role do ordinary men, or professional warriors for that matter, play in fighting to victory? Similar questions had arisen before around new weaponry (firearms, for example, were once seen as cowardly weapons), but in a war of large scale industry, mechanical weapons, extensive reporting, literate combatants, and a reading public, technology became critical as never before. In mechanizing the fight, the ship that surrounded these men with iron threatened their heroism, their self-image as warriors, even their lives. William Frederick Keeler, the *Monitor*'s paymaster, had a premonition while the ship was still under construction. Keeler pondered his new vessel's thick armor and novel machinery. "We were in no danger from shot or shell," he mused, "but thought the trip in her not quite safe." Keeler then posed a question that would forever color his reactions to the *Monitor;* he wondered whether "there isn't danger enough to give us glory."[6]

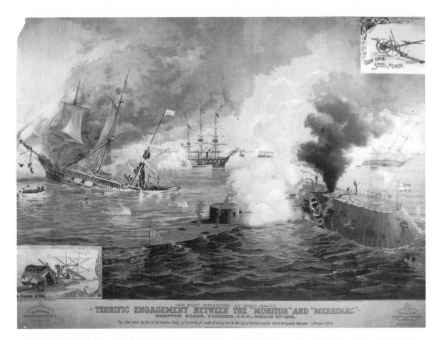

"The fight settled the fate of the 'Wooden Walls' of the World and taught all nations that the War Ship of the future must be—like the McCormick Harvester—a Machine of Steel," proclaimed this promotional poster in 1893, tying the war machine to civilian technology and the mechanization of agriculture. The *Monitor* symbolized American mechanical know-how, and military success suggested strength in commercial technology as well. Courtesy of Mariner's Museum, Newport News, Va.

The machine's status as an icon, nearly a hero itself, altered its users' sense of heroism. Remarkably, amid the stories and celebration, this new experience of war remains largely unexamined. Neither the *Monitor* crew nor the Union leadership could separate the symbolic, public aspect of the weapon from its military effectiveness. Hence, the *Monitor*'s story concerns not only the machine, but also the histories of expertise, experience, and representation that created it. Reexamining the *Monitor*'s history along these axes not only elucidates the technical changes of the nineteenth-century navy, but it also links those changes to general questions about technology and war. Ultimately, the Civil War was not thoroughly mechanized in the way later conflicts were; powered machinery appeared more in production and logistics than in battle. Nonetheless, the conflict raised questions that reappeared in the wars of the twentieth century, in which mechanization on a large scale did transform the face of battle. In those wars, as with the *Monitor*, the ap-

pearance of weapons comprised part of their effectiveness, alternately in-
spiring confidence, terror, and anxiety. Furthermore, new technologies—
whether tanks creaking across the Western Front in 1916, pressurized B-29
bombers over Japan in 1945, or automated missiles striking televised targets
in 1991—challenged accepted definitions of what it meant to be a warrior.
Cocooned machine operators, despite their risks, might still ask themselves
Keeler's question, whether "there isn't danger enough to give us glory." Indeed,
the Congressional Medal of Honor, which marks the summit of American
military bravery, has been awarded only twice since the Vietnam War, an unanticipated consequence of remote battles and unmanned weapons.[7]

These personal aspects of military technology are critical to understanding
modern warfare. They also engage broader questions of technology and culture: the status of the individual in a world of complex systems, the human
agency behind seemingly autonomous technological progress, and the nature
of technology as an aggregate of material artifacts and cultural representations
(itself a new idea in the *Monitor*'s day). War machines like the *Monitor* com-
press these themes, often subtle and diffuse in civilian life, into a compre-
hensible unity, a singular time and place where human effort concentrates
into struggles for dominance, survival, and attention. The *Monitor* story raises
questions that ultimately concern the fundamental tensions of technological
society.

The Irony of War in the Industrial Age

The American Civil War saw unprecedented use of industrial techniques
and resources for military operations. Submarines, torpedoes, reconnaissance
balloons, and of course, ironclad warships appeared as novel experiments.
Railroads, mass-produced rifles, and steam power changed the shape and
place of the battlefield. Photography and the telegraph brought the daily
conduct of the war to the public as never before. What began as a civil police
action, even a glorious lark, ended as industrialized slaughter on an entirely
new scale.

As it unfolded, this transformation altered the experience of soldiers. At
the war's outset, enthusiastic young men sought tests of individual skill and
bravery, ending either in victory or glorious death. In 1861 soldiers marched
to war "confident that the individual would remain the determinant of war's
course, that their personal goals would remain more important and chal-
lenging than any collective or organizational requirement."[8] The early war of
1861–62, bloody as it was, largely fulfilled these expectations. As the war
dragged on, however, it became something soldiers had not imagined when
they enlisted. Men burrowed into the earth and fought in trenches, much as
they would fifty years later in the fields of Belgium and France. Instead of

individual heroism, they found anonymity, stalemate, and impersonal annihilation, undermining the very values for which they were fighting. One
officer, a railroad engineer in civilian life, found at Petersburg in 1864 "a mode
of warfare strangely differing from the dashing cavalry service on the outposts.
Instead of the movement and excitement of rapid marches, sudden attacks,
and thrilling personal adventures, alternating with periods of gay social intercourse, I was now thrown into the very jaws of the grim death struggle in the
trenches of a vigorously besieged town, and was to begin a strange sort of warfare underground."[9] The components of heroism—duty, honor, manhood,
and courage—notions central to American men's experience of the Civil
War—were threatened by the war's industrialized tide.[10]

The Battle of Hampton Roads belongs to the earlier phase of the Civil War,
before the new face of conflict showed its muddy self. But the *Monitor* crew
felt its pinch as sharply as did soldiers in the trenches. There, subtle changes
accrued gradually, the effects of embedded systems of production, transpor-
tation, and supply. The environment on the *Monitor,* however, was overtly
mechanical. Itself the product of numerous systems (of invention, materials,
production, and armaments), the ship embodied and represented its networks
in a coherent unit, wheezing, snorting, and churning like the icon of the machine age it became. In this strange world, the crew lived underwater in a
sealed compartment. Breathing processed air and propelled by steam, they
were driven into battle by engineers, not by sailors. Living, sailing, and fighting on the *Monitor* reenacted, in a confined space, societal dilemmas about
skill and machinery. In the early-nineteenth-century civilian world, the arti-
san's skill yielded to machines and standardization as they transformed an
increasing number of human crafts.[11] Aboard the *Monitor,* mechanization
raised a similar question: What would be the fate of heroism—that martial
equivalent of artisinal pride—when men fight from an artificial world, protected behind iron plates?

Not only paymaster Keeler posed this question, but also one of the *Monitor's* many distinguished visitors. "How can an admiral condescend to go to
sea in an iron pot?" Nathaniel Hawthorne asked after visiting the ship. "All
the pomp and splendor of naval warfare are gone by." In the spring of 1862,
a few weeks after Hampton Roads, Hawthorne traveled from Massachusetts
to Virginia for a firsthand inspection of the war. Writing in the *Atlantic
Monthly,* he sensed the same tension that bothered Keeler. "Henceforth there
must come up a race of enginemen and smoke-blackened canoneers who will
hammer away at their enemies under the direction of a single pair of eyes;
and even heroism—so deadly a gripe is Science laying on our noble possibilities—will become a quality of very minor importance, when its possessor
cannot break through the iron crust of his own armament and give the world
a glimpse of it."[12] Hawthorne raised the question of whether there would be

heroes and tied it to social class: Who would heroes be? The "race of engine-men" and the "smoke-blackened canoneers" were not the usual bearers of naval glory, but new kinds of professionals and new kinds of workers—less akin to the strutting, Nelsonian heroes of 1812 than to rough mechanics and gritty factory operatives.

Another contemporary writer reformulated Hawthorne's questions in an explicit analogy between factories and battlefields. Herman Melville, in his Civil War poems, published in *Battle Pieces and Aspects of the War* (1866), explored machinery's impact on war:

> Warriors
> Are now but operatives; War's made
> Less grand than Peace.

("operative" being then a common term for a machine tender or factory worker).[13] Four poems from *Battle Pieces* take as their subjects the *Monitor* and the *Virginia*. Two, "The *Cumberland*" and "The *Temeraire*," share the standard public vision of the ironclads, the passing of the "wooden walls" and the nostalgia for the days of "navies old and oaken," a time in which Melville himself was at home. Two others, however, "In the Turret" and "A Utilitarian View of the *Monitor*'s Fight," address the social changes brought by the warship.[14]

Melville and Hawthorne echoed Keeler's personal observations but generalized them to human encounters with technology. Industrialized warfare seemed to enact what they had written about in novels and short stories: unswerving mechanism struggling with and overcoming organic spirit. In their earlier works, Hawthorne and Melville had explored the changing character of an agrarian America as it faced factories, railroads, and westward expansion. The nineteenth-century credo of progress dictated not only that democracy and humankind were perfectible through reason, but also that perfection could be hastened by human effort and industry. "Progress was conceived not simply to be part of a remote, ageless cycle, but something accessible, responsive to manipulation."[15]

Progress brought its own ironies, however. Industrialization ravaged the land as it enriched society. Factories ground down workers even while paying them good wages. Industrial war came laden with similar contradictions. By enrolling rationality and industry to kill, machines of death stressed the belief in progress and human perfectibility to its limits. Most of those who wrote about the *Monitor* dealt with this paradox simply by repressing it, overshadowing anxious personal narratives with talk of technological revolution. In contrast, Hawthorne and Melville, struggling to comprehend the nexus of technology and war, searched for language that would not collapse under the weight of its own irony.

History and the Experience of Technology

The theme of technological revolution has formed a critical part of the history of the *Monitor*. Even a casual bibliography would list hundreds of articles, memoirs, and books on the topic—beginning the day after the battle and continuing uninterrupted to the present. An inventor defends his creation against criticism. An officer writes to a superior about a successful operation. A sailor sends a letter to his wife about a harrowing experience. An old man authors an article to prove his heroism in a battle decades before. A historian traces origins of the military-industrial complex. Such accounts together comprise the history of the *Monitor*. Each retelling modified the history according to its particular time and place, but a standard, virtually canonical tale emerged, concentrating on the revolutionary nature of the machine. The Battle of Hampton Roads is the climactic event, the *Monitor's* sinking nine months later a tragic denouement.

Departing from the standard history, this book explores the experience of the *Monitor* and its representations. The *Monitor* did not revolutionize warfare. Rather, it redefined the relationship between people and machines in war. To characterize that change, one must examine the lived experience of battle—elusive, transitory, and difficult to document as it might be—and how it altered with the introduction of new technology. Such efforts pose no small challenge, for in contrast to the well-ordered world of commanders and strategy (and consequently, the historians who write about them), the battle of soldiers and sailors takes place in "a wildly unstable physical and emotional environment." More than twenty years ago, John Keegan called for historians to pay attention to this difficult subject, what he called "the face of battle"—referring both to the human visage of war and to the forward face of an army, where it actually contacts the enemy. Of course, the experience of war has always been a source of anecdotes, stories, and personal memoirs, but only relatively recently have historians begun to focus on soldiers' battle experience as a serious topic of historical investigation.[16]

Civil War historians such as James McPherson and Gerald Linderman have laid a foundation for studying the role of experience in the *Monitor's* history. They traced the persistence and erosion of soldiers' patriotism, duty, honor, and loyalty to comrades in the face of radical changes at the front. These and similar studies enrich our knowledge of the conditions of frontline combat in the Civil War and their ideological and psychological underpinnings.[17] But none address technology, beyond the way it changed the conduct of battles on a broad scale (an understandable focus given that such studies have tended to focus on army rather than navy experience), although scholars have recently begun to examine the experience of civilian technology in early industrial America.[18] Keegan himself paid little attention to the effects of mechanization, what we might call "the face of technology." What becomes of people when the face of battle becomes that of a machine? The history of the *Monitor* begins to answer this question, but only if we expand "experience" to include the effects of machinery on living conditions, daily routines, interpersonal relationships, relations with family and the public, and participants' expectations of battle.

We have, of course, no direct access to any person's experience in a time so long past. We can only approximate and reconstruct it from the traces it left behind, in this case primarily letters, diaries, and other forms of writing. Many crewmen left such traces, but paymaster Keeler's are the most detailed and thoughtful. His letters to his wife, Anna, are well known and appear in most histories of the *Monitor.* Those histories invariably present him as an observer, though, rather than an actor, as a member of the audience rather than a member of the cast. Rarely do they treat Keeler on his own terms as a player in a historical drama or consider his experience worthy of historical study in its own right. Here I consider Keeler a *participant* as well as a *witness* and attend not only to the events he observed but also to the fabric of life aboard the *Monitor* and how he made sense of its novel conditions. Numerous other forms of documentation, particularly those of John Ericsson, navy officials, and others of the *Monitor* crew, overlap and extend Keeler's observations, but his provide a clarity and an immediacy unmatched in other accounts of the ironclad. Furthermore, reading Keeler alongside Hawthorne and Melville imbues him with a significance broader than the Battle of Hampton Roads or even the advent of ironclad warships. The three men's writings place the *Monitor* at the intersection of technology and war, where invention collides with annihilation.

Opening the Iron Box

Granted, library shelves strain with volumes on military technologies. Nearly all, however, separate the development of weapons from their application and use. Such stories tend to have simplistic views of invention—in the inevitable march of progress, once a device was invented, it was easily accepted; those who opposed it are accused of "resistance to change." This determinism treats technology as an external, autonomous force that impacts upon culture.[19] The standard history of the *Monitor* took this form: John Ericsson, the heroic inventor common in nineteenth-century stories of technology, had to struggle to gain acceptance for his wonderful new machine against forces of conservatism and self-interest, though he was eventually proven right by success at Hampton Roads. This story has overlooked the fact that, then and now, the success or failure of an invention is itself the outcome of debate and controversy.

Using the *Monitor*'s history, we can hope to understand such contention and its outcomes of success or failure and critically examine the mechanisms of technological change. This approach is colloquially known as "opening the black box"—that is, closely examining the design, construction, and use of technology. All are social and political, as well as technical activities. Historians have not frequently opened the black box of military technology, but there have been significant forays. Donald Mackenzie has argued, for example, that in our own day engineering groups emphasized the accuracy of ballistic missiles—previously considered a purely objective measure of performance—to bolster their own investments in certain guidance techniques. The "natural trajectory" of guidance technology, ever increasing its accuracy as if propelled by a force of its own logic, is but one indication of the technology's progress—others believed that cost or reliability, not accuracy, was the relevant axis along which to measure progress.[20] The natural trajectory is inevitably the history of the victors, and it suggests a narrow set of reasons for success or failure. In contrast, historians now aim to see weapons not as the results of inevitable, incremental progress but as components of social and institutional, as well as technical, spheres. Military enterprises, in this view, are not simply passive users of technology but are sources of patronage, management techniques, and intellectual traditions that shape the course of invention.[21]

Nearly all histories of the *Monitor* follow the natural-trajectory narrative, arguing for the ship's significance as the initiator of the inevitable march toward the modern steel battleship. But if we open up the *Monitor*'s opaque skin and peer inside, we see that no technical innovation caused an instant revolution in naval warfare. Rather, the *Monitor* succeeded because it answered questions and solved problems of naval warfare and industrial weaponry *and* because it had powerful backers and a successful public demonstration. Delving into the debates that surrounded the *Monitor* and characterizing the problems it was attempting to solve, we find disagreement on even the meaning of success for an ironclad. Numerous criteria could be proposed: victory in a single battle, a long and distinguished career, production of a line of similar vessels, influence on future designs. Furthermore, so many variables enter the equation for performance of a warship (speed, firepower, and thickness of armor, to name but a few) that design decisions become judgments about the relative importance of competing variables. What seem at first to be purely technical debates emerge, under close inspection, as contests over the very nature of navies and warfare.

John Ericsson, the designer of the *Monitor*, was well aware of these debates. He argued that the *Monitor* successfully solved open technical problems in naval warfare. The technical questions had social implications, however, and he was frequently challenged by others with equally plausible solutions. How ships were going to be made affected who would design them, who would build them, and who would fight in them. Everyone was an interested party. To support their positions, these parties brought to bear different types of evidence of success. Included, of course, were the technical capabilities of the machine, but its public presentation, the actions of its inventor, and even the geographic conditions of Hampton Roads were also crucial parts of the picture. That the ship was generally considered successful was a measure of Ericsson's (and his allies') success in silencing his critics and rivals. And succeed they did for a while: most of the subsequent Civil War ironclads followed the "monitor" type. But in the long term, the monitor type did not succeed as a stable solution to problems of naval warfare—thus highlighting the local, temporary nature of the original *Monitor's* triumph.

This said, opening the black box of the *Monitor* does not entail contrasting its success in the public sphere with another measure within a larger technical history. Put another way, this book does not seek the *Monitor's real* place in the technical history in contrast with its cultural representations, but rather treats the two as intertwined, indeed inseparable. John Ericsson, the *Monitor's* inventor, knew that the appearance of a revolutionary weapon could be as significant as its performance in battle and that performance in battle adds as much weight to technical arguments as do engineering measures such as design studies or controlled experiments. In addition, those who fought and lived aboard the *Monitor* were keenly aware of their symbolic role in military politics, and that role materially affected their experience aboard the radical vessel, as well as their ability to fight.

Chapter One

Revising the
Revolution

1815–1861

About an hour's train ride south of London, one can visit today the Portsmouth Naval Shipyard, the centuries-old home of the Royal Navy, now an impressive museum. There one can see the remains of the *Mary Rose,* the pride of Henry VIII that sank in 1545 and has been recently recovered by underwater archaeologists. Next door to the *Mary Rose* is the well-preserved *Victory,* Lord Nelson's flagship during his triumph over the French and Spanish fleets at Trafalgar in 1805. The point where Nelson fell to a sniper's bullet is well marked on the main deck, and the site of his death deep within the hull has been turned into a shrine. Although dedicated to the human hero, the shrine might well mark the passing of a technology as well, for the massive *Victory,* weighty with timber frames, represents the apotheosis of the "wooden walls," Trafalgar the last great encounter between these sail-adorned machines.

Not far from the *Mary Rose* and the *Victory* lies another antique warship, beautifully restored. The *Warrior* was built in 1860, half a dynamic century after Trafalgar; it represented the high state of the engineering art when the *Monitor* appeared two years later. From afar, the *Warrior* appears to be a traditional sailing vessel, though its lines are sleek and straight compared to the *Victory*'s chubby girth. Upon closer inspection, however, more modern features are evident: black sides dotted with rivets fastening iron plate, two funnels atop massive steam boilers, a screw propeller hanging forward of the rudder, and steam-driven ventilators to keep the interior air above atmospheric pressure. The iron armor, four and a half inches thick and backed by eleven inches of teak, surrounds a boxy citadel amidships that protects the engines and the main armament.

Whereas the *Victory* represents the perfection of a type and the climax of a historical period, the *Warrior* appears, despite its physical grace, to be an awkward experiment. Although a seagoing ship, the *Warrior's* coal-hungry boilers kept it close to home, patrolling the channel to offset the naval challenge from France.[1] The pointed knee-bow was an archaism, structurally unnecessary in an iron ship but aesthetically connecting the *Warrior* to the era of wood. The vessel had so many novel features that it effectively made itself obsolete, as its new ideas developed further in other construction projects. Rather than constituting a radical departure in naval design, the *Warrior* represented a nodal, transitional point at which designers and constructors experimented and discussed (with varying degrees of vitriol) the best combinations of new technologies. One historian labeled as "the era of uncertainty" not the period leading up to the *Warrior* but the period that followed it.[2] Though no revolutionary invention, the *Warrior* did represent Britain's first, careful foray into armored, seagoing steam warships, and therein lies its enduring elegance and interest. In contrast to the radical, experimental *Monitor,* the *Warrior* embodied state technology, the smart, thought-through solution of a secure officialdom.

Making a Revolution

As the *Warrior* demonstrates, when the *Monitor* and the *Virginia* met in 1862, they did not break into a world unaware of the prospects of steam and iron armor. Rather, they continued an international conversation that had engaged the European powers for several decades. Naval architecture was not experiencing one revolution, but several simultaneous ones. Cladding a ship with armor was no simple matter of adding plate to keep out shells. It was part of a calculus with many variables, including technical issues of materials, propulsion, engines, and structure. It encompassed military considerations of speed, force structure, tactics, and grand strategy. Nor was it immune from the politics of industrial endowments, patronage, national finance, and public opinion.

In 1862, though locked in a desperate civil war with a proximate foe, Americans celebrated the *Monitor's* international effects. European navies, with their heavy, intimidating investments in sailing ships, had the most to lose from the advent of ironclads. Isaac Newton, chief engineer of the *Monitor,* repeated a popular mantra when he credited his vessel with "nearly annihilating the offensive power of France & England" and predicted that the ship would "disarm naval powers."[3] In contemporary Europe, popular opinion mirrored the Americans' pride with anxiety. The *Times* of London reported, "It must now be looked upon as proved that wooden vessels go to sure and speedy defeat whenever they venture into action against an iron-plated adversary." American Henry Adams, living in London in 1862, reported the British mood: "The *Warrior* and their other new iron ships, are no better than wood, nor can any shot-proof sea-going vessel be made . . . within three weeks, they [the British] find their wooden navy, their iron navy, and their costly guns all antiquated and useless."[4] One recent history declared, "After March 8–9, 1862, no major naval power in the world was safe, even against a second- or third-rate nation, if that lesser power's fleet could boast an ironclad."[5]

News of Hampton Roads had enough impact in Britain that debates arose in Parliament over the effects of the ironclads and whether the system of armored forts then under construction should be abandoned. Nonetheless, confidence in the country's naval and technical power remained solid. All anticipated the change from wood to iron, but members argued (to a resounding "Hear, Hear") that "England, with its great wealth, its mechanical appliances, and ample supply of coal and iron, could not only provide for the preservation but the maintenance [of the dominance] it had hitherto enjoyed."[6] British experts and policymakers never felt as strongly about the new technology as their American counterparts. Indeed, American news of the collapse of wooden walls rarely took account of the existing state of naval technology in 1862. Europeans had long been grappling with problems of ironclad warships.

The restored HMS *Warrior* in a recent photo at Portsmouth, England. Built in 1860, it represented the innovative, yet conservative, British solution that the *Monitor* challenged.

Their experience had shown, in fact, that the application of new methods of engineering to naval warfare involved much more than simple calculations. Experiments could effectively close debate in the engineering and scientific worlds, but they did not end conversation in the military world. Experience in battle, no matter how chaotic, carried more weight than carefully controlled results on the proving ground. This tension, between the methodical precision of engineering and the dramatic experience of war, underlay the evolution of naval technology in the mid–nineteenth century.

"What the Naval Constructors Already Knew"

In 1860 the major naval powers of the world, Britain and France, had to protect colonial empires with far-flung holdings and global trade routes. Both countries were at work on numerous ironclad warships, long-range seagoing vessels that could sail and fight in heavy seas. Not one, but at least seven major innovations affected these endeavors: exploding shells, steam power, screw propulsion, iron construction, armor, rifled ordnance, and slow-burning powder.[7] Each bore heavily on the others; all were under way in the decades preceding the Civil War.

Before the construction of the *Monitor,* both Britain and France had recognized the passing of the wooden ship. Napoleon III and his innovative technical officers, particularly Henri Paixhans and Dupuy de Lôme, inaugurated the era of armored warships with a series of experiments and inventions, culminating with the construction in 1859 of the armor-clad *Gloire.* The British followed a close second with the *Warrior* of 1860. The two countries laid down more ironclads in the eighteen months *before* Hampton Roads than they did in the two years *after.* By 1861 Britain had suspended new construction of wooden war vessels altogether, but Britain and France retained large, effective wooden fleets. Because they could not sail, much less fight, in anything but calm, coastal waters or rivers, the American ironclads *Monitor* and *Virginia* were suited only for defense of the essentially island nation. As James Phinney Baxter wrote in 1933,

> The combats of March 8 and 9 [1862] symbolized the passing of the old fleets and the coming of the new. *Symbol they were, and not the cause of the revolution in naval technology,* for they did not initiate the great revolution in naval architecture, they crowned it. They taught the man in the street what the naval constructors already knew: that shell guns sounded the doom of the wooden navies of the world. On the chief problem confronting the naval constructors of Europe—the best design for *seagoing* ironclads—these battles threw little light.[8]

Although the American ironclads achieved success in the public eye, they did not achieve it in the technical realm. According to Baxter, the *Monitor*'s significance lies not in its machinery but in its role as a cultural icon. "The significance of the *Monitor*," as the argument goes, "relates primarily to its 'mythic' qualities," that is, the ship's impact on American consciousness, rather than to a supposed technological revolution. After Hampton Roads, a "*Monitor* Craze" swept the Union. Editorials, cartoons, songs, and other expressions of popular culture used the icon of the *Monitor* to signify American technological strength. And historians helped the *Monitor* become "a symbol of American ingenuity and know-how, the progenitor of the modern battleship."[9] Baxter distinguishes between such "symbols" and the actual "causes" of the revolution in naval architecture.

Beyond Symbols and Causes

To be sure, the *Monitor* fused the symbolic power of the machine with national goals and military power into a formidable icon. Ships have long symbolized technical achievement, national pride, maritime power, and a host of other human accomplishments (witness the enduring appeal of the *Victory* on display in Portsmouth or the *Constitution* in Boston). In the decades preceding the Civil War, railroads began to displace ships as symbols of the mechanical age; Americans began to project their fears and aspirations onto graceful, breathy locomotives. The *Monitor* drew upon this mechanical iconography but combined it with the visceral stirrings ships had inspired for centuries. In fact, before 1862 the most famous American warship was the *Constitution,* nicknamed "old ironsides," a fact not lost on the ironclads' promoters (one of the *Monitor*'s competitors became the *New Ironsides*).

Cultural historian Michael Smith helps make sense of the relationship between the symbols and the causes the *Monitor* embodied. In the twentieth century, Smith observes, "each new product of technology was really two: the device itself and the image of the device in the mind of the consumer or enemy," a phenomenon he calls technological display.[10] The *Apollo* space program, for example, was both a technical exercise to reach the moon and a program to generate the public image of American superiority—the moon rocket was simultaneously a complex machine and a symbol of the people. Long before *Apollo,* the *Monitor* demonstrated that the image of a machine could be as potent as the machine itself.

Yet machinery is not so easily disentangled from its representations. Observers of and participants in the *Monitor*'s history made no neat distinctions between images and devices, between symbols and causes. For John Ericsson, the *Monitor* represented American know-how because it embodied American know-how. The same goes for the crew, the politicians, and the public. Their attitudes reflect the fact that technology, right down to armor plate and turret bearings, is part of culture, a cultural expression. Participants do not paste symbols onto a previously existing industrial base; technical reality does not exist independent of cultural significance. Each influences the other, to the point where distinctions between them become difficult to maintain. Making technologies is a symbolic activity, and making symbols is a technical activity. Both constitute what we call technology.

The very notion of technology was a recent invention in the *Monitor*'s day. It appeared in its modern form in 1828, underwent its infancy during the Civil War, and was used infrequently before the end of the century. The idea eventually became current to capture the sense that modern technical enterprises—with their significant financial, political, and symbolic dimensions—were inadequately described by traditional terms such as "mechanic arts." "The habit of separating the practical and the fine arts," writes Leo Marx, "had served to ratify a set of invidious distinctions between things and ideas, the physical and the mental, the mundane and the ideal, body and soul, female and male, making and thinking."[11] A machine like the *Monitor* did not fit neatly into these dichotomies, nor into the neat distinction between symbols and causes. In that sense the story of the *Monitor* itself helps to tell the history of the *idea* of technology, since it dates from a period when technical enterprises began to comprise arrays of experts, state support, and public imagery, in addition to machine-building.

Thus, the *Monitor* was an early technology as well as a machine (although it was not usually described that way in its time). The public *Monitor* and the technical *Monitor* constantly played off one another—technical success depended on successful representations, and public recognition depended on successful technical performance. Often, ship and icon were not in harmony. Participants in the *Monitor*'s history frequently commented on the dissonance between the machine and its representations. There was a wide space between its performance as theater and its performance as a machine.[12] All machines have technical qualities in tension with their images and expectations, but rarely is the contrast so extreme. The *Monitor,* more than the numerous other ironclads of its day, is particularly illustrative of the issues surrounding the historical conception of technology.

The historical intricacy of the *Monitor*'s development, then, not only shows the influence of social factors on design but also effaces the very distinction between inside (technical) and outside (cultural). Nations build public statements when they build warships; inventors build symbols as well as machines; individuals incorporate public perceptions into their experience of new machinery; popular opinion affects decisions to develop new techniques. Understanding the *Monitor* as one, not two, refines the notion of technology as a social and technical enterprise involving things, events, observers, and documents. The struggles surrounding naval warfare in the mid–nineteenth century were birth pangs of modern technology—wherein large, complex machines resulted from vast systems of skill, natural resources, labor, and finance and performed roles simultaneously private, symbolic, political, and military.

Revising the Revolution: Britain and France

Consider, from this perspective, the history of naval technology in the decades before the Civil War. Stable solutions—those that would have lasting, rather than temporary, impact—required more than straightforward engineering logic and procedure. Inventing the future of naval warfare involved debate, not only among engineers but also including admirals, captains, politicians, shipwrights, sailors, the public, and the press, all from several different countries. During the decades preceding Hampton Roads, these groups in Britain and France engaged in what today we would call a naval arms race. In the international conversation, "both countries believed, and perhaps even encouraged, exaggerated reports of the other's potential" (a dynamic familiar to post–Cold War Americans).[13] For those arguing a position, numerous types of evidence contributed to differing positions in the debate: foreign intelligence (often rumors and hearsay), experiments, publications, results of sea trials, shipbuilding programs, and especially combat experience. They did not all carry equal weight.

Today, the notion that military professions embrace technology seems unproblematic; a significant number of military leaders are trained as engineers, in fact, and most accept the doctrine that advanced weaponry is a key to winning wars. In the early nineteenth century, however, this idea was far from universal—military arts did not include engineering and often required no mathematics. The military industrial worldview that emphasizes uniformity, novelty, and calculation began with a cadre of expert artillery officers in revolutionary France. Gradually, their techniques found their way into the ancient profession of naval ship design and construction, using calculation and quantitative physical theory to predict the behavior of ships and hence predicate designs. The advent of ironclad vessels roughly coincided with the rise of what we know today as "naval architecture," the engineering approach to ship design. But rather than a mathematically determined optimum, every warship was a series of tradeoffs, among speed, weight of armor, number of guns, endurance, and cost.[14] Engineering technique helped stabilize the axes of these parameters, but only experience could establish priorities among them for a particular mission.

In the early nineteenth century, military engineering shared much with its commercial counterpart, but at least one fact set it apart: success in battle carried the greatest weight of all evidence. Victory could enshrine a technology considered inferior in peacetime. Defeat could undo years of experiments. A battle-tested solution, even if proven inferior by other standards, could take decades to unseat by technical argument. Naval battles do not take place under controlled conditions. Numerous other factors besides technical capability affect the outcome: the skill of commanders, the training and motivation of crews, the weather, and sheer circumstance, to name but a few. Even the outcomes themselves are debatable, because victory and defeat, like success and failure, are categories open to interpretation. Herein lies the paradox of applying engineering methods to military technology: the single most persuasive piece of evidence for success or failure hovers on the edge of chaos and bears little resemblance to a controlled experiment. As Karl Von Clausewitz, the preeminent theoretician of nineteenth-century land warfare, wrote, "the absolute, the mathematical, as it is called, nowhere finds any sure basis in the calculations in the Art of War . . . from the outset there is a play of possibilities, probabilities, good and bad luck, which spreads about with all the coarse and fine threads of its web, and makes War of all branches of human activity the most like a gambling game."[15] Nonetheless, their very chaotic nature gives battles great power as deciders of technical suitability. Despite all the uncertainty, the argument goes, if a new technology has performed successfully in battle, it must be ready for general application. During the period 1815–60, a time of relative peace in Europe, those battles that did occur loomed particularly important.

A warship is a social machine. Its design has to suit not only the workings of its own crew, but also the characteristics of its intended enemies. Numerous elements enter the design calculus for a warship, including industrial endowments, financial capacity, political maneuvering, technical skills, the properties of materials, and basic physics; also entering the equation are the actions (real and imagined) of others, particularly foreign navies. Consider the example of the strength of a particular kind of armor, a seemingly straightforward matter. In the mid–nineteenth century, armor was tested by firing experimental shots into material samples on a firing range. Such trials depended on assumptions that were difficult to validate or even to identify: the size of the enemy's guns, the type of projectiles they would fire, and the expected range of an encounter. In 1860 answers to these and numerous questions like them were just emerging from confusion.

Similarly, the famous transition "from sail to steam" occurred only after decades of uncertainty and interdependent developments. In the early nineteenth century, Europeans embraced steam rather quickly, but largely for auxiliary vessels and tugs rather than the large line-of-battle ships. In the antebellum decades in the United States, commercial shipping took great advantage of steam power, usually using paddle wheels to transmit energy from engine to water. Such an arrangement, though acceptable for commercial cruising, presented several critical handicaps for a warship: the wheels occupied significant broadside space that could otherwise be dedicated to armament, they were vulnerably exposed to the enemy, and in their typical direct-drive configuration they forced the steam plant to be located above the water line, also a vulnerable position. To become acceptable for major warships, steam would require another invention: the screw propeller. Propellers allowed the machinery to be located belowdecks and the propulsive gear under water. But John Ericsson found that such logic was by no means self-evident to the parties involved (see chapter 2); the Admiralty was reluctant to replace a well-known technology with an unproved one.[16]

In their uncertainty, the Royal Navy took cues from civilian designers. In the 1840s the navy built a series of vessels and conducted trials influenced by the screw-driven merchant ship *Great Britain,* which was designed by the legendary engineer Isambard Kingdom Brunel. Undertaken with Brunel's guidance, these experiments improved the form of propellers and the hulls required to accommodate them, as well as the understanding of the device's action—all necessary before the screw could compete with paddle wheels. A further series of trials in 1845, with the demonstration ship *Rattler,* settled the technical questions for an admiralty board, because the *Rattler* won a tug-of-war with a paddle vessel. Technical data themselves were insufficient; the Admiralty needed this spectacle to close the issue and win broad support. "It is fairly certain . . . that the Board had decided in favor of the screw before this tug-of-war was held. In all probability, it was a spectacular public relations exercise, designed to convert remaining doubters."[17] The British introduced steam-driven, screw-propelled battleships in 1849, and by the end of the following decade the new propulsion method had largely taken over from paddles.

Changes in propulsion combined with changes in armor and ordnance—particularly the introduction of guns that fired fused, exploding shells instead of solid shot. French artillery engineer Henri Paixhans made this old idea practicable in the 1820s, and by the end of the following decade the French, British, Russian, and American navies had adopted shell guns, but as secondary armaments, not main batteries. Shell guns became primary only when combat seemed to demonstrate their value. At Sinop in 1853, during the Crimean War, the Russians utterly destroyed the Turkish fleet with exploding shells. Numerous observers saw in the encounter an unequivocal demonstration that shell guns spelled the end of wooden navies. Expert analysts, however, were not so persuaded by the drama. John Dahlgren, the designer of the *Monitor*'s guns, analyzed the battle of Sinop in his famous 1856 treatise, *Shells and Shell Guns.* He concluded that the results of the Crimean War were "so far qualified by circumstances as to be unavoidably partial in their character." The skirmish at Sinop, writes engineer-historian David K. Brown, "had effects out of all proportion to its operational significance." Their impressive showing notwithstanding, shell guns still had problems with reliability, safety, and rate of fire.[18]

Despite these uncertainties, the combat experience at Sinop overrode previous technical arguments about shell guns. The French, convinced by the episode of the power of shell guns, soon fielded a set of light-draft "floating batteries" with a few heavy guns and armor to protect them against shell fire. These devices proved effective in both offense and defense, helping to capture the Russian fort at Kinburn on the Black Sea. But again, their success was far from conclusive: they had resisted only relatively light shot fired from a long range (24 lb. shot at 1,000 yards).[19] Nevertheless, the batteries of the Crimean war—ungainly, poorly ventilated, and difficult to control—made an important symbolic point more powerfully than years of experiments could have: shell guns and armor threatened the status quo in warship design.

New ideas about armor in turn led to uncertainties in construction. Iron armor added significant weight, which had implications for a ship's structure and raised questions about whether a new warship should have a wooden hull with iron armor (literally ironclad) or an entirely iron hull. Advocates of iron for hulls argued that it increased strength and durability, reduced the risk of fire, and allowed the construction of watertight compartments. Other arguments had to do with propulsion: wooden hulls were ill suited to support the stresses of large, screw-driven warships, so adopting the propeller gave impetus to iron construction. Iron ships would also finally solve Britain's centuries-old problem of timber supply, although this fact was curiously not emphasized as an argument for iron ships.[20] Opponents of iron construction found it contrary to nature to build ships out of a substance that would not float on its own. They argued that iron hulls would wreak havoc with magnetic compasses. Iron also had severe problems with fouling by marine organisms in salt water: even in the American Civil War, iron ships (including the *Monitor*) were reduced to half their original speed after several months because of marine growth. Perhaps most important, it was not at all clear that iron would resist enemy fire any better than wood. Before the Crimean War, between 1846 and 1851, a series of firing-range tests in England, well planned and scientifically conducted, showed iron hulls, under varying conditions, to be *more* vulnerable and dangerous than wood. Britain subsequently abandoned iron construction, although it did not abandon armor on wooden hulls; France and the United States soon followed.[21]

This hiatus in iron construction, though initiated by controlled experiments, ended when combat experience overrode the evidence. The performance of French floating batteries in the Crimea prodded the Europeans back into seriously contemplating iron construction. In 1859 the French launched the *Gloire*, with a wooden hull (stiffened with iron) and armor plate. France intended the *Gloire*, as the Americans intended the *Monitor* sixteen months later, to overturn British superiority by a bold step forward. Also like the *Monitor*, the *Gloire* struck fear into the British public; it initiated an invasion scare that demanded a response.[22] The Admiralty, on the defensive, argued that they had been waiting for just this eventuality. Surveyor of the Navy Sir Baldwin Walker wrote, in an oft-noted passage,

> Although I have frequently stated that it is not in the interest of Great Britain— possessing as she does so large a navy—to adopt any important change in the construction of ships of war which might have the effect of rendering necessary the introduction of a new class of very costly vessels, until such a course is forced upon her by the adoption by foreign powers of formidable ships of a novel character requiring similar ships to cope with them, yet it then becomes a matter not only of expediency, but of absolute necessity . . . This time has arrived.[23]

In a response as symbolic as it was practical, Britain answered the *Gloire* in kind with the *Warrior* in 1860, followed soon after by the *Black Prince,* the first seagoing ironclads with iron hulls. Walker's goal was not to change the character of the navy—he still considered wooden frames clad with iron to be the best solution—but rather to supplement the force with experimental ironclads and meet the French one-for-one in their latest challenge.[24] The *Warrior* was built by private contractors (with an Admiralty design) and not, as was the custom with wooden ships, in government yards, which lacked appropriate facilities for processing iron. Walker's policy may or may not have been the best policy for the navy, but the *Warrior* did stabilize decades of experience and debate into a workable design. In its time the most expensive warship ever built, it represented the state of the art in seagoing ironclads when the *Monitor* was laid down in 1861. It was this careful, conservative solution that the *Monitor* threatened.

In the 1850s the Royal Navy, secure in its design, engineering, and industrial capacity, had no need for drama; it was pleased to allow others to go first and take the risks, a policy that invited much criticism. The British response to the *Gloire* reminds one of Walter MacDougall's observation about the launch of *Sputnik*—the Soviets stunned the world with *Sputnik* not because the American space program was weak, but rather because the American program, being strong and comprehensive, did not see the need for a dramatic first.[25] As with *Sputnik,* this strategy can work except for one factor: the potent, unpredictable, symbolic power of machines.

The American Navy and the New Expert Officers

In the 1840s and 1850s the United States had an ambivalent relationship to the developments in naval technology. The U.S. Navy had achieved even less consensus on the future of naval warfare than the British. American invention continued apace, but the navy did not follow up a number of major experiments with building programs. Robert Fulton's *Demologos,* for example, achieved notoriety in 1814 as the first steam-powered warship, but it was never operationally deployed, and it spawned no line of similar vessels, except the unremarkable *Fulton II* in 1837.[26] Five years later the American Congress let a contract for the "Stevens Battery" to New Jersey inventor Robert Stevens, but the project, never completed, was canceled after Stevens's death and many failed attempts at completion. Senior officers and administrators also saw few advantages in the new propulsion. The *Princeton,* built by Robert Stockton and John Ericsson, introduced the screw propeller to warship construction in 1842, but it too ended up an isolated experiment. Numerous commercial ships employed this innovation before the navy built further screw-propelled warships a decade later.

The U.S. Navy simply did not have the support in Congress that the Royal Navy had in Parliament. With no far-flung colonial empire to defend, America relied on a naval strategy of coast and harbor defenses and had little incentive to build a seagoing navy in peacetime.[27] It did, however, build up a conventional steam-powered force in the 1840s and 1850s (the last sailing-only vessel built for the navy was completed in 1855), part of which saw service in the Mexican war. The most prominent of these ships, the *Merrimack* class of screw-driven frigates, entered service in 1855, and twelve screw-powered sloops were authorized in 1858. The *Merrimack* itself, which in modified form would face the *Monitor,* was commissioned in Boston on June 15, 1855, as a screw-propelled frigate with full sailing rig.[28] American shipbuilders, however, notably lagged behind the British in iron construction, largely for economic reasons: lumber was inexpensive in the United States, iron expensive—the reverse of Britain's situation.[29] In 1860 the United States did not have a single ironclad warship. Even in the Naval Appropriations Act of 1861, Congress appropriated *no* money for ironclad ship construction, despite approving more than a million dollars for steam-powered sloops.[30] So hesitant was the American attitude that British ironclad opponents used the abstinence of the progressive nation as an argument against the new technology.

This state of affairs led to the now-familiar charges of resistance to change and conservatism regarding technology in the antebellum navy. Serious problems did plague the service. War heroes occupied the few upper ranks in a hierarchy where promotion was based entirely on seniority (defending against a naval aristocracy, Congress allowed no rank higher than captain), and senior officers kept their posts for life. The antebellum navy developed two tiers: a few senior officers locking up the top positions, and numerous junior officers who could go decades without promotion. This system led to accusations of "old-fogyism" during the Civil War, charges John Ericsson would use to great advantage.

Slow movement in construction and promotion, however, masked significant developments as the social underpinnings of technological change grew quietly within the U.S. Navy. The "American military enlightenment" occurred in the 1830s and 1840s: a sense of military professionalism gradually combined with technical specialization to foster a class of expert officers.[31] The antebellum navy built an international reputation for science, exploration, and survey through its support of the nation's commercial interests along the coasts and abroad. Junior officers, with little chance for traditional promotion, turned to scientific and technical expertise to distinguish themselves. These men, born in the new century, became known as the "steam generation" of naval officers and advocated not only new machinery but educational and administrative reform as well. Many did scientific service in the U.S. Coast Survey, among the earliest government research establishments, and they brought their skills to their naval commands. John Dahlgren, for example, left the Coast Survey for the Washington Navy Yard, where he used surveying techniques to evaluate firing tests in his pioneering work on naval ordnance. The *Monitor* carried his famous soda-bottle-shaped guns known as "Dahlgrens." These smoothbores, considerably lighter and more powerful than their predecessors, had nine-, ten-, eleven-, thirteen-, and eventually fifteen-inch bores; they were the most powerful naval guns of the war.[32] Matthew Fontaine Maury, at the Naval Observatory, collected data to track whaling grounds, charted currents and weather conditions, and contributed to fundamental oceanography and hydrography. Maury also joined others, such as Samuel Francis DuPont, in successfully calling for reform of the navy's administrative structure. The Naval Academy at Annapolis was established in 1845, and Franklin Buchanan, who later commanded the *Virginia,* became its first superintendent. The academy became a full four-year school in 1850 and laid the educational groundwork for a professionally trained corps of career naval officers. Other steam-generation officers included Matthew Calbraith Perry, John Rodgers, David Dixon Porter, David Farragut, and Raphael Semmes.[33] These men enjoyed the prime of their careers during the Civil War, although they did not all fight on the same side.

Of the new expert officers, the most institutionalized and the most contentious were the engineers. An act of Congress in 1842 created the navy's Engineering Corps, consisting of the ranks engineer-in-chief, chief engineer, and assistant engineer. The men in these ranks not only supervised and operated the steam plants of ships at sea; they also designed the engines for ships under construction. Most came from the extensive machine-shop culture that developed around railroads and manufacturing in the antebellum decades, trained entirely by practice and apprenticeship and then certified by examinations within the navy. Yet mechanical skill, despite its growing importance, did not have the prestige of shiphandling or tactics. Engineering "staff" officers suffered the disdain of mainstream "line" officers (the men who commanded and fought the ships) who considered them lowly mechanics.[34] Frank Bennett, writing a history of the Engineering Corps in 1898, captures the professional anxiety that lingered among the engineers as they toiled away in hot, dirty engine rooms while line officers fought gloriously on deck:

> It has been written that it is difficult to become sentimental about the engineer. This idea is born of the belief that he deals with material things and takes no part in the glorious possibilities of war or in the victories that are won from storms. This theory is absolutely false; . . . an admiration for the sea and those who face its dangers on the part of those who never go to sea has made of the sailor's existence a picturesque ideal that has become an article of faith with landsmen. And this faith excludes the new type of seaman—the man of the engine and boiler rooms—from any share in the romance of the sea because he faces dangers of another kind and performs his duty in another atmosphere, though equally exposed to the dangers that are peculiar to life afloat. When some poet with a clearer vision and a willingness to enter an untrodden field shall appear and sing the song of steam it will be a revelation for the multitude; for there is music and romance and poetry as well as the embodiment of power about the mechanisms that drive the great ships of today.[35]

American steam engineers, unlike the French artillerists, did not have a broadly rationalizing mission; they sought not to be engineers of war in general, but only of steam machinery. They would have to struggle even for that privilege.

One engineer in particular epitomized the triumphs and anxieties of the young Engineering Corps. Benjamin Franklin Isherwood came to the navy a technically inclined young man with experience in railroads and lighthouses. He worked in a private firm, the Novelty Ironworks of New York City (which later built the *Monitor*'s turret), in order to learn his way around marine engines and secure a naval appointment. In 1847, after serving on a number of navy steamers, Isherwood was appointed first assistant engineer. He became chief engineer two years later. Soon he was working in the engineer-in-chief's office designing machinery.[36] By 1860 Isherwood separated himself from the tinkerers and engine tenders who preceded him: he conducted a series of detailed experiments on an old steamer, measuring power and efficiency under controlled conditions. Isherwood's conclusions on the expansive power of steam were controversial, but he collected a great deal of data and in 1863 produced a two-volume work, *Experimental Researches in Steam Engineering*, that became a standard text on the subject. The book analyzed the power and efficiency of the *Merrimack* class of screw frigates and included detailed calculations of boilers and pumps.[37] Isherwood did much to link practical problems in engineering with the emerging science of thermodynamics. He stressed inductive reasoning from data to physical principles, methodical experimenting, and publication of results. "Sagacious comparison of the results of accurate observation is our sole teacher," he wrote. "The first care of the investigator must be to acquire precise experimental data from which to proceed, instead of commencing with imaginary assumptions that can only end in a vain parade of misdirected skill."[38] Isherwood disdained mathematics and theory; his extreme empiricism reflected his practical, nonscholastic roots.

Isherwood's career epitomized the concerns faced by new technical professionals in the Civil War navy. They did not have the prestige of college degrees but had learned their profession at the nation's informal engineering schools: the canals, railroads, steamships, and machine shops that proliferated in the 1830s and 1840s. A few of them rose to scholarly attainment, as Isherwood did with his publications and as John Dahlgren had in the field of ordnance (Dahlgren studied with the French artillerists, as did several others of the steam generation). In 1861, buoyed by the support of his mentor, John Lenthall, chief of the navy's Bureau of Construction, Equipment, and Repair, Isherwood became engineer in chief of the U.S. Navy.[39] He was 39 years old.

As head of a critical and growing organization, Isherwood played a central role in the Civil War navy. He became chief of the Bureau of Steam Engineering at its inception in 1862, among the most powerful staff jobs in the service. From this position he undertook the Herculean task of supervising the design, construction, operation, and maintenance of all of the navy's propulsion machinery. He tirelessly promoted the professional, institutional, and educational causes of the Engineering Corps, but during the Civil War it still struggled for legitimacy. An 1859 order conferred "naval rank" on the engineers, the equivalent of commander, lieutenant, master, or midshipman of varying grades, but the men had "no authority to exercise military command," even over the men on their own ships.[40] Their position was never secure. Isherwood frequently found himself the target of acrimony, always criticized, always second-guessed by line officers who felt that the hard-pressed new bureau did not always serve their needs. Private engineers (frequently with congressional connections) also derided Isherwood. *Monitor* builder John Ericsson became an outspoken critic (Isherwood's *Experimental Researches* had a chapter on the *Monitor,* showing from experimental data that it contained superior boilers but inefficient engines). Americans may have welcomed technical change, but they did not always welcome the new professions that accompanied the inventions.

Building the Civil War Navy

The Civil War began with a naval action, the firing on Fort Sumter in April 1861. When President Lincoln called up the troops and imposed the blockade of Southern ports as part of the Anaconda Plan to strangle the South, the blockade's enforcement became the dominant naval activity of the war. But the conflict found the U.S. Navy a small and largely outdated force. Of its ninety vessels, only twenty-one were powered by steam (in comparison to six hundred steam vessels added during the Civil War), and many of those were scattered in distant seas. Secession also damaged and demoralized the human organization—nearly 25 percent of the 1,554 officers "went south," through resignation or dismissal, during the crisis of 1861, including Matthew Fontaine Maury and Franklin Buchanan, commandant of the Washington Navy Yard, who later commanded the *Virginia* at Hampton Roads. Senator Stephen Mallory, chairman of the Senate Naval Affairs committee, also joined the Confederacy.[41] The Confederate navy, in worse shape than that of the Union, had essentially no warships at its inception. Mallory, secretary of the navy for the Confederacy, had followed closely the debates and developments surrounding iron, armor, and steam in Europe. The Confederacy, at Mallory's urging, quickly initiated an ironclad program to make up for the South's inferior numbers with invulnerable vessels.

The South's ambitions in ironclad technology would have been stillborn without the windfall of a major industrial facility. Soon after Fort Sumter, they captured the U.S. Navy's Gosport Navy Yard at Norfolk, Virginia, a critical strategic asset. With it they took numerous ships, guns, and, most important, fabrication and dry-dock facilities. Secretary of the Navy Gideon Welles, a Connecticut newspaperman turned skilled administrator, sent Benjamin Isherwood to Norfolk to try to rescue the valuable frigate *Merrimack,* but the engineer could not overcome the hindrance of the incompetent yard commander. The powerful warship had to be burned to the waterline to prevent its capture by the rebels. The loss of the navy yard remained a controversial embarrassment to Welles, and his support of the ironclad program stemmed in part from a desire to compensate for that fiasco.[42]

Welles began to build the federal navy, but he faced considerable challenges, material, political, and bureaucratic. Throughout his tenure he struggled with Lincoln's secretary of state, William Seward, powerful congressmen, and the navy itself. As one observer put it, "It would seem that the one subject with which the direction of naval affairs had never concerned itself was the subject of making war."[43] To begin assembling a wartime force, Welles initiated significant building programs in the spring and summer of 1861. By December the navy had purchased 136 vessels, including 26 side-wheel steamers and 43 screw steamers, and had under construction 14 screw sloops, 23 gunboats, and 12 side-wheel steamers. Between March and December 1861, the organization grew from less than eight thousand to twenty-two thousand men; it would expand to more than 600 ships and fifty thousand men before the war's end. This mobilization in effect solidified the commitment to steam. Welles boosted the authority of Isherwood and Dahlgren when he declared, "No sailing vessels have been ordered to be built, for steam, as well as heavy ordnance, has become an indispensable element of the most efficient naval power."[44]

This massive effort not only supported steam but also brought new attention to the question of ironclads. In July of 1861, when Gideon Welles submitted his report to Congress, he suggested that a board be convened to evaluate ironclad vessels. "Other governments, and particular France and England, have made it a special object in connection with naval improvements,"[45] Welles wrote, initiating the ironclad program less to pursue a technological imperative than out of watchful caution regarding other nations. Several events in addition to the general buildup contributed to this belated, although urgent, official interest in ironclads. Government indifference to new technology had not been lost on the press: in March of 1861, for example, the *Philadelphia Examiner* opined:

> It is a curious fact that the United States . . . should be charged with being behind the age . . . It is more than probable that, and without much further delay, if we intend to have a national naval force worthy of our power and pretensions, we shall have to follow suit, and build iron-cased vessels, as France and England have done, and are doing. Before the end of this year, France will have eight and England six such vessels. How many are we to have?

Expert advice supported the view, in the form of a report that John Dahlgren submitted to Welles on European ironclad construction projects.[46] Furthermore, the Confederacy used a rudimentary floating battery in the attack on Fort Sumter. Soon after the Norfolk navy yard fell, reports began drifting north that Confederate armorers had raised the burned hulk of the *Merrimack* and were cladding it with iron. Welles's recommendation for the institution of the Ironclad Board partly reflected his concern about what the South was rumored to be doing in its Norfolk facility.

Congress responded to Welles's recommendation by incorporating it nearly verbatim into an August authorization, ordering the secretary "to appoint a board of three skillful naval officers to investigate the plans and specifications that may be submitted for the construction or completing of iron or steel-clad steamships or steam batteries" and appropriating $1.5 million for the purpose.[47] Significant as these actions were, however, neither represented an endorsement of ironclad warships. Welles noted, "The period is, perhaps, not one best adapted to heavy expenditures by way of experiment," but he argued that the funds he requested would allow construction of such an experiment before deciding on larger appropriations for operational vessels.[48]

Welles found his bureau chiefs (including Dahlgren and Isherwood), the navy's highest technical experts, to be wary of the ironclad plan. Turning instead to line officers, he appointed to the new board Joseph Smith, 71, chief of the Bureau of Yards and Docks, Hiram Paulding, 64, who had fought in the War of 1812 and had just retired from naval service when the war started, and Charles H. Davis, 54, a man of mathematical skill and scientific accomplishment who had worked at the Coast Survey and had prepared the naval almanac. The board received offers from British shipbuilders to provide ironclads, but it chose instead to foster domestic production. The Department of the Navy published in the major East Coast dailies a request for proposals to build "Iron-Clad Steam Vessels . . . for sea or river service."[49] Interested constructors were asked to provide descriptions and drawings of their proposed vessels, as well as armor and machinery, and they were to include not only schedules for completion, but also a guarantee for "proper execution of the contract."[50]

The Ironclad Board was not charged with supporting ironclad technology but rather with evaluating it. Granted, the appropriation of $1.5 million itself gave a significant boost to the legitimacy of ironclads, and three major construction projects would certainly advance the state of the art. Nevertheless, terms like "experiment," and "evaluation" pervade these documents. The caution is understandable given that members of the board were not experts, or even necessarily current in the ironclad debates of the day. Smith and Paulding represented the older, traditional sailing navy. Davis, though younger and scientifically literate, had no expertise in construction or steam machinery. The board's final report disclaims its own knowledge. "Distrustful of our ability to discharge this duty," they wrote, "we approach the subject with diffidence, having no experience and but scanty knowledge in this branch of naval architecture . . . it is very likely that some of our conclusions may prove erroneous." They doubted the practicality of armored oceangoing cruisers, questioned the effectiveness of shell guns and rifled ordnance, and noted that "high authorities" in England still differed about whether iron or wood construction was best.[51]

Whatever their unwritten motivations, both Welles's initial recommendation and the report of the board cited their primary driver as the building programs of foreign navies. "The construction of iron-clad steamships of war is now zealously claiming the attention of foreign naval powers . . . As cruising vessels, we are skeptical as to their advantages and ultimate adoption. But whilst other nations are endeavoring to perfect them, we must not remain idle."[52] For engineers and constructors, the issues surrounding ironclad warships hinged on the fine points of armor, design, and weapons, but these would not convince this nontechnical group of experienced commanders. The Ironclad Board responded to international events; it examined a technology about which it was uncertain, if not downright skeptical. Only when

Welles personally assumed the risk and responsibility for building the ships did the board agree to act. Ironclad warships had much to prove.

Nevertheless, the board did believe armored ships could do well as light-draft vessels in harbors and rivers, although they (correctly) did not think ironclads could destroy masonry fortresses. The board received sixteen proposals, most of them from companies in the northeastern United States but one from Ohio (one was from the clipper ship builder Donald McKay). Estimated costs for the vessels ranged from $32,000 to $1.5 million. Designs varied considerably, one submission even proposing a rubber-clad vessel. Of these proposals, the board chose three: one by Merrick & Sons of Philadelphia, "the most practicable one for heavy armor"; one by C. S. Bushnell and Company of New Haven, Connecticut; and one by John Ericsson of New York, for a "floating battery." Contracts were recommended for all three, each with "a guarantee and forfeiture in case of failure." These contracts would produce the Union's first ironclad warships: the *New Ironsides,* the *Galena,* and the *Monitor,* respectively. The egg-shaped *Galena,* 210 feet long, had rounded sides to deflect shot and novel armor plate fastened by rails. The huge *New Ironsides,* in contrast, was designed most like a European ironclad, 249 feet long with a submerged ram, angled sides, and an armor belt.[53] Of the three, however, the board still described only one, the *Monitor,* as an "experiment."[54]

John Ericsson had managed to convince the Ironclad Board to accept his radical, untried idea. According to Ericsson's own account, it was a pure technical demonstration: the board called him in, he explained his design and presented buoyancy and stability calculations with absolute clarity and certainty, and his proposal was accepted immediately. "In less than an hour," Ericsson wrote, "I succeeded in demonstrating to the satisfaction of the board appointed by President Lincoln that the design was thoroughly practical and based on sound theory."[55] This version coincides with the larger mythology of the *Monitor* as the product of an individual mind, that of the great inventor John Ericsson. The legend treats the individual genius in Washington and the novel warship in Hampton Roads in analogous terms: the radical outsiders, arriving with technical superiority to settle the great debates of naval construction against great conservatism and resistance.

Ericsson did go to Washington. Documents show, however, that the naval board did not credit his explanation as much as he believed and reported. After the presentation, they expressed hesitation over Ericsson's battery: "We are somewhat apprehensive that her properties for sea are not such as a sea-going vessel should possess." The board approved the proposal only with Welles's encouragement and assurances. Furthermore, although Ericsson was the person most strongly identified with the proposal, he had significant allies, each of whom brought political, financial, and industrial resources to bear on the proposition. John Ericsson's radical, heroic role contained a mixture of technical and political activity, which are difficult to disentangle.

The *Monitor* did not arise in isolation but participated in an ongoing series of developments in naval technology. On the international scene, these developments sought seagoing ironclads that could defend the global empires run by the European powers. Builders and navies experimented with new configurations of weapons, structures, and propulsion. The *Monitor* settled these debates only locally, if at all, for it satisfied only a limited set of American requirements; the following era of uncertainty saw numerous further experiments in warship design and configuration. Yet the *Monitor* did have significant importance in the war of images and politics that accompanied expensive new naval technologies. War in the industrialized world was not comprised of technical solutions alone; it also included interwoven systems of politics, public perception, and international rivalry. Warships as cultural products embodied a new notion of technology, wherein complex machines emerged from extensive social and industrial networks and had to satisfy numerous and diverse interests. To achieve success for his invention, John Ericsson had to accommodate or overcome numerous interested parties: a sophisticated European establishment that believed it could respond to and not initiate innovation, American naval experts who had their own solutions to naval problems, and an Ironclad Board skeptical of the machine's potential. To contend with these forces, Ericsson realized he would need two allies, less technical but equally powerful: politics and public representation.

Chapter Two

Building a Ship, Speaking Success

The term "monitor" is often thought, naturally enough, to derive from its function of guarding or monitoring a blockade, a river, or a harbor. It actually describes the political impact Ericsson hoped the vessel would have. Writing to Gustavus Fox, the inventor emphasized the public perception of the technology in the United States and Britain as much as its utility in fighting a war:

> The impregnable and aggressive character of this vessel will *admonish* the leaders of the Southern Rebellion that the batteries on the banks of their rivers will no longer present barriers to the entrance of union forces . . . but there are other leaders who will also be startled and admonished . . . Downing St. will hardly view with indifference this last Yankee notion, the monitor. To the Lords of the Admiralty the new craft will be a *monitor,* suggesting doubts as to the propriety of completing those four steel-clad ships at three and a half millions a-piece. On these and many similar grounds I propose to name the new battery *Monitor.*[1]

The *Monitor* had to communicate several overlapping messages, each to different groups and each with different meanings. For the navy, the *Monitor* had to satisfy a contract so that its builders would be paid and would be asked to build more ships of the same type. To the public in America and in Europe, particularly Britain, the *Monitor* had to convey the strength of American industry and ingenuity that could be quickly brought to bear as military force. Finally, and most obviously, to the Confederacy and to its navy, the *Monitor* had to represent the industrial power of the Union, against which innovations and barricades would be futile.

For the ship to succeed it would need to affect a broad and sophisticated audience. The builders of the *Monitor* purposefully constructed a symbol even while they fabricated a machine.

John Ericsson and the Theme of Resistance

INCEPTION OF THE MONITOR

Ericsson's Preparation for His Great Work.—His Struggles with Professional Jealousy.—Dealings with the Navy Department Previous to 1861.—Presents Two Subaquatic Systems of Attack to the Emperor of the French.—History of Armored Vessels.—Outbreak of the Civil War.—Prompt Action of the Confederate Authorities.—Ericsson Offers His Services to President Lincoln.—Is Called to Washington.— Dramatic Interview with the Board on Armor-Clads.—The Monitor Ordered.
Chapter heading from William Conant Church, *Life of John Ericsson*

The historical Ericsson was born simultaneously with the historical *Monitor*, for John Ericsson made his own history. Late in his life, when asked by a museum director to donate personal records, Ericsson replied, "I have already destroyed upward of one thousand drawings, and numerous models, to prevent posterity from supposing that my knowledge was as imperfect as said relics would indicate."[2] The documents he destroyed, diaries and personal papers, mostly dated from before the construction of the *Monitor*. Ericsson thus identified his own biography with that of the famous ship, both effectively beginning in 1861.

Ericsson left no memoir. Much of what we know about his early career, indeed about much of his life, comes from an adoring biography written in 1890 at Ericsson's request by his friend William Conant Church. Church, a journalist turned editor, published the *Army and Navy Journal,* a conservative gazette of military affairs. During and after the war, Church and his journal were deeply involved in supporting Ericsson's position in controversies over warship design and steam engineering. Indeed, most of Church's editorial material on naval shipbuilding issued directly from Ericsson's pen. The chapter heading quoted above suggests the tone of Church's biography, the heroic mode pioneered by Samuel Smiles in his *Lives of the Engineers*.[3] It chronicles how "Nature and Opportunity combined their forces to produce the great engineer," a man constantly "at war . . . with received opinions on engineering subjects," and it jibes with Ericsson's own accounts.[4]

It would be a mistake, however, to dismiss Church's biography because of its proximity to Ericsson and his debates. When understood in light of its conditions of production, and in conjunction with other documentation, the Church biography provides an important window into the inventor's self-fashioning (in addition to accurately, although selectively, reprinting numerous important documents). A consistent story emerges. The visionary inventor clashes with stodgy, conservative, and entrenched bureaucracies. Resistance to change by the inventor's opponents appears as a constant theme. Ericsson has several frustrating experiences, at first in England and then with the U.S. government. He withdraws his public spirit into private ventures but is coaxed out of isolation by the crisis of war, when he selflessly submits his new design for the good of the nation. Even after the success of the *Monitor* at Hampton Roads, those who hold the reins of power still doubt the inventor's genius. The history, as told by Church and Ericsson, is one of conservatism, often the result of political interests, that the inventor must overcome in order to convince others of the technical superiority of his invention, a superiority that would be obvious but for irrational, complicating factors.

Yet indicting historical actors for failing to see the future, as do Church, Ericsson, and other *Monitor* historians, amounts only to accusing them of not knowing what we know today. Irrational opposition surely comprises part of the story—the navy of the 1860s had its share of problems—but it makes a frail crutch on which to support an inventor's heroic biography or the dramatic tale of an instant and obvious naval revolution. Telling a different kind of story makes more sense of the human actors in the drama and requires discarding less of their activity as irrational. A more balanced history hinges on the simple recognition that successful inventions are understood as successful only in hindsight; when they are introduced, no one knows whether they will succeed or fail. At every point success or failure is open to question, indeed hotly debated. As with the debates over naval warfare explored in chapter 1, several types of evidence are brought to bear in determining the success of a technology: technical logic, practical demonstration, commercial demand, expert opinion, combat experience. Approaching the history this way does not require debunking Ericsson's heroic legend, nor even that of the *Monitor's* success; rather it entails a critical evaluation of what made them so. Instead of asking who opposed the invention, we must frame the question positively: Who supported it? What convinced them it was successful? What strategies did the inventor use to convince others—his partners, the navy, the public, historians—that his invention was successful? How were opponents either converted or silenced? What finally settled the debate?

Invention and Frustration

John Ericsson's career before the *Monitor* demonstrated to him that technical superiority alone, or technical superiority as he defined it, would not be sufficient to ensure the success of an invention. Ericsson was born in Sweden, the son of a miner, in 1803. His father, "a great admirer of Pohlem," the Swedish inventor and model maker, worked on the country's great engineering endeavor, the Gotha Canal.[5] Two sons, the older Nils and the younger

John, grew up in the engineering culture surrounding the project. Under the tutelage of a German engineer-officer, John displayed a particular talent for engineering drawing and began working on the canal in his early teens, acquiring, at least according to Church's account, the status of protégé. Ericsson the engineer would always be the draftsman and the calculator, more than the tinkerer or the constructor. After seven years on the canal project run by the navy, the intense, stocky Ericsson entered the Swedish army, where he studied artillery and land surveying, two disciplines central to the military science of the time. Ericsson's drafting skills advanced him quickly. Soon the king appointed him lieutenant and put him to work drawing maps of the empire. After a failed romance and the birth of an illegitimate son, the 24-year-old Ericsson left Sweden for England with borrowed money and an idea for a new engine driven by flame instead of steam. This idea evolved into Ericsson's hot-air engine, the elegant invention on which he worked his entire life.

Once in England, Ericsson could not build his new engine; instead, he joined an engineering firm headed by John Braithwaite and soon became the junior partner of Braithwaite and Ericsson. Here the Swede patented and built several new machines, largely for mining and drainage, and continued his dream of replacing steam with a new motive force. To pay the bills, however, Braithwaite and Ericsson improved existing steam engines and built refrigerators and coolers for breweries and distilleries.[6] In 1827 Ericsson built his first marine engine, for the *Victory,* an experimental craft for an arctic expedition to find a northwest passage. The trip failed and the ship sank. Ericsson was blamed for the disaster, but he asserted that his client had deceived him about the vessel's purpose by asking for an experimental war steamer rather than an arctic explorer. During these years Ericsson also made a number of early steam-powered fire engines that also, in Ericsson's view, succeeded technically but seemed too novel to be accepted by the existing fire brigades.[7] "In his contest with the London Fire Brigade," writes Church, "Ericsson appears to have had his first introduction to the official inertia and prejudice he was destined to become further acquainted with during his long career of invention."[8] Phrased differently, Ericsson began to realize that in addition to construction, successful inventions required persuasion.

In a foray into railway engineering, Ericsson learned the importance of public demonstration in establishing technical superiority. Like the domain of naval warfare, the railroad world was engaged in fundamental debates, in this case whether to apply power to railways with mobile engines (locomotives) or fixed ones (as used in mines and factories). In an attempt to settle the debate, the Liverpool and Manchester Railroad offered a prize for a locomotive that could pull a load on rails at ten miles per hour, among other specifications of size and weight, in a set of trials in October 1829 at Rainhill. The father and son George and Robert Stephenson, experienced builders of industrial engines, entered a radical design dubbed the *Rocket*. Ericsson, working frantically at the last minute, built an entry he called the *Novelty*. Other entries competed but were disqualified early, although one, the *Sans Pareil*, ran the full trial as a serious competitor. Over the course of several days, before crowds numbered in the thousands, the *Novelty* ran faster than the *Rocket* on a straightaway, unloaded (31.9 versus 24 miles per hour). In a distance trial with a simulated load, however, the *Rocket* ran steadily above twenty miles per hour. The *Novelty* ran smoothly: "We can say for ourselves that we never enjoyed anything in the way of travelling more," wrote a journalist who went along for the ride. "The velocity was such that we could scarcely distinguish objects as we passed by them, the motion was so steady and equable that we could manage not only to read, but write." Nevertheless, the *Novelty* repeatedly broke down with mechanical problems; finally Ericsson withdrew his entry.[9] In the spectacle, Ericsson certainly drew the attention of the crowd and the praise of the engineering press, but the *Rocket* took the prize. Whatever the theoretical advantages of Ericsson's engine, the practicalities of construction kept it from performing for the world on the appointed occasion. Today's engineers will recognize the situation: Ericsson had failed the demo.

Despite losing the Rainhill competition, during the 1830s Ericsson's career advanced steadily. He produced a host of inventions, collecting thirty patents in his ten years in England, and continued work on his hot-air caloric engine. This engine was truly Ericsson's lifework, but despite local technical successes, it never succeeded as a technology of wide importance. As in the area of locomotive design, considerable disagreement existed over the possibility and the practicality of hot-air engines, particularly in the absence of a coherent theory of heat and work. Practically, though, air engines required exceedingly high temperatures for operation, which made them heavy and took a toll on the metals and lubricants of the day. Ericsson lost the priority of patenting to Robert Stirling, the Scot whose name is associated with hot-air engines to this day.

In still another episode, Ericsson conducted a successful demonstration but failed to sell his invention—leading him to suspect powerful social forces arrayed against the introduction of new mechanisms. As part of his work of improving the efficiency of steam engines, Ericsson developed a screw propeller for steamships. Potentially more efficient and less vulnerable to enemy fire than paddle wheels, the screw propeller allowed the steam machinery of a warship to be located below the water line, where it, like the propulsion gear, would be protected. He built an experimental ship, the *Francis B. Ogden*, in 1837, named after his American patron, to showcase his new propeller. That summer, he demonstrated the vessel to a group of the Admiralty's highest officers—including William Symonds, the surveyor of the navy—by towing an Admiralty barge in a day trip up the Thames from London. Ericsson thought the demonstration a success—the ship navigated freely at greater than ten knots—but the Admiralty expressed no further interest, deeming the idea impracticable and inefficient. They had seen steering problems in screw-driven ships before, and they thought towing the barge was a ruse to disguise the problem. Ericsson also failed to get the English patent on the screw propeller. In England patents go to the first applicant, in this case Francis Pettit Smith, rather than the first inventor, as in America. Thus, despite a number of inventions, a wealth of experience, and a respected name among engineers, Ericsson found his years in England frustrating. All his major ideas were stillborn, and Ericsson's firm failed in 1837. That same year, amid hard economic times in England, he declared bankruptcy and found himself in debtor's prison.

Ericsson in America

Despite Ericsson's repeated misfortunes, there were two Americans in England who had great interest in the engineer, now 34 years old. Accompanying Ericsson and the Admiralty on the propeller's test run on the Thames had been Robert F. Stockton, an American colleague of Ericsson's patron Francis Ogden. Stockton, from an old New Jersey family, was at the time actively engaged in building canals and railroads in his home state. With Stockton's support, Ericsson built another screw steamer, this one of iron, the *Robert F. Stockton*, which ran successfully in 1838. The following May, the *Stockton* crossed the Atlantic, and in November 1839 Ericsson himself boarded a steamer and headed for New York, leaving behind his wife, Amelia, whom he had married three years before. Although she followed her husband to New York, Amelia Ericsson soon returned to England. John Ericsson sustained her financially for the rest of his life, but they never met again.

Before leaving England, Ericsson and several associates designed a wrought-iron naval gun and had it built in Liverpool, with Stockton's financial support. To strengthen the longitudinal iron bars that composed the barrel, Ericsson shrunk a band of iron around the breech. The following year, in response to a proposal by Stockton, the U.S. Navy ordered Ericsson to build a steam-powered warship. This ship became the *Princeton*, with a propeller and machinery of Ericsson design, the first screw-propelled steam warship. Ericsson always considered its unique semicylinder engines among his best work.[10] In addition to a standard battery, Stockton and Ericsson mounted two novel guns on the *Princeton*. One, the wrought-iron gun shipped over from England, was dubbed the Oregon, in response to a current dispute between the United States and Britain. Stockton ordered a second, the Peacemaker, in New York, which was built under Ericsson's supervision but lacked the shrunken band for strength. Launched in 1843, the *Princeton* proved to be a vessel of "exceptional performance and remarkable efficiency."[11]

Stockton saw the ship as the harbinger of an expanded and technologically revitalized navy. But in his zeal to convert the Washington powerful to his view, Stockton brought tragedy on the *Princeton*. On February 28, 1844, the Peacemaker (the gun without Ericsson's strengthening iron band) exploded amid a coterie of distinguished guests. Several important men were killed, including Secretary of State Abel P. Upshur and Secretary of the Navy Thomas W. Gilmer. President Tyler might well have been among the casualties, but he was belowdecks at the time. This time the demo had failed fatally, with legal and technical consequences. A court of inquiry found the episode to be an accident, and Tyler supported Stockton in further endeavors. The Franklin Institute conducted a technical investigation, however, which found the cause of the failure to be faulty construction techniques; the navy refused to order any further vessels of the *Princeton* type or any wrought-iron guns—dismissing Ericsson's working embodiment with Stockton's faulty one.[12] Ericsson, for his part, had been snubbed by Stockton, literally left waiting on the dock, on the way to the fatal demonstration. He believed Stockton sought credit for the many Ericsson novelties on the vessel, a charge with some merit. Stockton never paid Ericsson for his work, and the inventor refused to defend him in the aftermath of the tragedy.

Ericsson thereafter had little involvement with military work in the 1840s and 1850s (although his few surviving papers do show he proposed several ideas to both branches of the service).[13] He formed a productive alliance with the Phoenix Foundry of New York and designed and built numerous steam engines and machinery for both marine and terrestrial applications. Ericsson even acquired financial backing for a full-size demonstration of his hot-air or "caloric" engines and built the 260-foot-long *Ericsson,* launched with much fanfare in 1854. The ship was initially received with great praise, but it garnered no long-term support and sank ingloriously in a storm in April of 1854. It was salvaged, but new steam engines replaced the ruined caloric ones.[14] Ericsson turned his attention to small caloric pumping engines, of which he sold more than a thousand in the 1850s.

Of Ericsson's difficult early career, William Conant Church wrote, "The reception, no less than the conception, of new ideas necessitates evolution, and this is a weary world for those who see much beyond their fellows."[15] Told this way, the story portrays the authorized Ericsson: the unrecognized genius overcome by social forces beyond his control—such as English hostility to new inventions. Undoubtedly, some nineteenth-century people found new inventions threatening. But Ericsson's heroic narrative is simply not supported by evidence. Victorian England, the home of such technical polymaths as John Scott Russell, Robert and George Stephenson, Thomas Telford, and Isambard Kingdom Brunel, deeply valued mechanical contributions. It celebrated engineering accomplishments from bridges, ships, and railways to public health systems. The Crystal Palace exhibition in 1851 showcased a host of technologies, from the iron and glass palace itself to cranes, machine tools, and marine engines, and even foreign ideas such as the American Colt revolver were imported. Similarly, as we have seen, the British Admiralty was itself struggling with numerous naval revolutions. They too embraced technological change, albeit conservatively. Ericsson's claim of resistance to change in Victorian society was groundless. Rather, what he learned in England was that success required more than simply his own judgment of an idea's technical merits. It had to include the judgment of numerous other parties as well. The *Princeton* fiasco proved even further, though in the negative sense, how profoundly those perceptions could affect new machinery, especially in the arenas of state sponsorship and war. Appearances count; demonstrations convince; nationality inspires; politics gets things done.

Building Alliances

In 1861, when the Ironclad Board issued its advertisement for ironclad proposals, John Ericsson was a successful professional engineer in New York, although one lacking in public recognition. Despite the *Princeton* setback, naval warfare had been on his mind. In 1854 he submitted a proposal to Napoleon III for a "new system of naval attack," to be used with dramatic effect in the Crimean War. The French never supported Ericsson's idea (they never even responded). The inventor hoped they would aim the weapon at Russia, the enemy of his homeland Sweden, and at England, the enemy of his genius.[16] No proof of Ericsson's proposal exists except for his own account; James Phinney Baxter exhaustively searched the French archives and turned up no evidence of Ericsson's submission. The inventor later wrote, "I never communicated my plan to any one before submitting it to emperor Napoleon. I imagined then I had a very valuable idea and kept the secret accordingly."[17] Ericsson believed that his ironclad would provide a clearly superior solution to the problem of naval warfare, if only he could overcome the resistance to change. From its inception, however, the *Monitor* project had to overcome both caution and competition.

In the weeks after Fort Sumter in the spring of 1861, before the Ironclad Board issued its request, a number of shipbuilders lobbied the navy for contracts. Ericsson was not among them, but Cornelius Bushnell was.[18] Bushnell, a young Connecticut industrialist and an organizer of the Union Pacific Railroad Company, went to Washington to find support for his railroad projects. At a meeting in the Willard Hotel, his friend from Connecticut, Navy

Secretary Gideon Welles, showed Bushnell the draft bill for an ironclad program, as yet unapproved. When Welles took it to the Senate Naval Affairs
Committee, Bushnell persuaded Senator James Grimes of Iowa of the potential value of ironclads as harbor defenses. Grimes took up the cause and ush-
ered the draft through the committee. Congress subsequently approved the
bill authorizing$1.5 million for ironclad construction. The president signed
the bill on August 2.[19]

Bushnell received a contract for the *Galena,* his own plan for an ironclad,
with the assistance of Samuel Pook, a Boston naval constructor. At the sug-
gestion of his colleague Cornelius Delamater, Bushnell asked Ericsson to review Pook's design for buoyancy and stability. Delamater's ironworks in New
York City was the largest manufacturer of marine steam engines in the United
States, and Delamater had a long-standing professional relationship and personal friendship with Ericsson. Delamater's company sold Ericsson's inventions, including caloric engines, and Ericsson employed Delamater facilities
for his work (Delamater had built the boilers, the propeller, and the blowers
for the *Princeton* and the *Ericsson*). Ericsson reported that Bushnell's craft
would be "not only sufficiently stable but what sailors term stiff." He also
mentioned his own idea for a novel warship, a low-freeboard iron raft with a
single revolving turret.[20] Much impressed, Bushnell became Ericsson's first
convert; he formed an alliance with Ericsson, boasting to the navy that he
had "secured the best talent in the country as I will show you when we come
to Washington."[21] He began lobbying Gideon Welles and the Ironclad Board
for the Ericsson proposal.

To strengthen his case, Bushnell sought the cooperation of two Albany iron
makers, John Griswold and John Winslow. Both had strong Washington connections, including their friend Secretary of State William Seward and a business partner, Erastus Corning, in Congress.[22] Bushnell offered Griswold and
Winslow a half-interest in the project, in return for their political, financial,
and industrial help. "I selected these Gentlemen," Bushnell later stated, "because of their large acquaintance with Government authorities, and because
I already had business relations with them, having contracted with them to
plate the *Galena*."[23] The deal worked: the two men secured a letter from
Seward and an audience with President Lincoln. With Bushnell as an ally,
Ericsson no longer had to play the outsider.

Lincoln thought the plan intriguing, but Bushnell found the Department
of the Navy skeptical of "another Ericsson failure." One member of the Ironclad Board, Captain Davis, the most scientifically literate of the group, firmly
opposed the idea. "Take the little thing home and worship it," he told Bushnell, "as it would not be idolatry because it was made in the image of nothing
in the heaven above or the earth below or the waters under the earth." Bushnell pleaded with Ericsson to come to Washington to make the case himself.

Bushnell's memoir describes Ericsson as bitter over his disappointment with the *Princeton* and wishing to have nothing to do with Washington. Ericsson may simply not have wished to take time from preparing drawings for the new vessel.[24] In any event, in mid-September, Ericsson did go to Washington. He met with the Ironclad Board and presented his stability calculations, whereupon, according to Ericsson's recollection, they became thoroughly convinced.[25] Gideon Welles gave Ericsson a verbal approval that day, and work began at once. On October 4 a contract was signed for an experimental "Iron Clad Shot-Proof Steam Battery." Ericsson and his backers were to deliver the vessel "complete ready for service in all respects excepting guns, ammunition, coal and stores" within one hundred days, for a price of $275,000 (it also needlessly required that they deliver spars and rigging, a provision the builders ignored).[26] The signatures of five men appeared on the contract: Welles, Bushnell, Ericsson, Winslow, and Griswold. Winslow put up the bulk of the money; Griswold supplied the armor; Bushnell secured a bond for his portion; Ericsson put up no money but did all the design and supervised construction.[27] Each had a quarter stake in any profits from the project.

Ericsson believed his technical arguments had overcome all opposition on the Ironclad Board, but in fact he had not been entirely persuasive.[28] When Bushnell hurried back to Washington to sign the contract, he found the board less sanguine than Ericsson had reported. Joseph Smith, the senior member of the board, wrote to Ericsson, "The board reported favorably on your proposition for an ironclad gun boat, but—as there seems to be some deficiencies in the specifications, as some changes may be suggested and a guarantee required, you had better come on and see to the drawing of a contract if we can mutually agree."[29] Similarly, the board recommended that the ship be built as "an experiment" if it could be done in one hundred days. In light of their doubts as to the vessel's stability, they demanded "a guarantee and forfeiture [of payments] in case of failure in any of the properties and points of the vessel as proposed."[30] Such clauses were not unusual in the contracts being let for standard vessels at the time; in fact they had been key to the navy's success in contracting privately built steamships before the war.[31] Typically, however, the contracts were contingent only on delivery, whereas the *Monitor* contract depended on performance under fire for its completion. This idea seems to have actually come from Ericsson's partner Winslow. Ericsson unhappily accepted it, lest his impregnable confidence lose its force. "It is hardly necessary for me to say," he wrote to Smith of the Ironclad Board, "that I deem your decision to test the 'Impregnable' battery under the enemy's fire, before accepting, perfectly reasonable and proper, if the structure cannot stand this test, then it is, indeed, worthless."[32] Nevertheless, the inventor considered it a concession, perhaps an insult to his own certainty of success, but he knew the new vessel required efforts on many fronts to succeed. "But for my desire to retain Mr. Winslow in the enterprise, on account of his relations with certain members of the administration," he wrote to Commodore Smith, acknowledging the importance of Winslow's political help, "I would never have accepted his amendment of the original contract."[33]

A Floating Locomotive

Ericsson, who expressed himself through drawings, conceived the *Monitor* as an aggregate of clean, geometrical shapes controlled by mechanical contrivances. What he produced was a bizarre craft that looked nothing like a traditional ship. The vessel, 172 feet long and 41 feet wide, incorporated two separate hulls. One, a flat iron raft, formed the deck (Ericsson called it the "upper hull"). The other, a flat-bottomed iron cradle, hung under the raft below the waterline (there was no true keel). An armor belt made of iron with pine and oak backing on the upper hull protected the lower hull from shot, shell, and ramming. The upper hull thus overhung the lower by several feet on each side and fore and aft, protecting the propeller and the anchor well. Buoyancy depended on a watertight seal where the two hulls joined, accomplished by vertical stanchions, brackets, and rivets (problems with this seal and the overhang of the upper hull plagued the monitors throughout their operational lifetimes and may have been responsible for the *Monitor*'s loss). The unusual "double trunk" vibrating-lever steam engine had a single cylinder, forty inches in diameter, which drove two pistons. It was designed to deliver four hundred horsepower to the four-bladed propeller of nine feet in diameter, although Isherwood measured only eighty-two under experimental conditions.[34]

All machinery, crew spaces, and stores were located belowdecks in the lower hull. Legend holds that the ship included fifty "patentable inventions," although that number is based only on a casual estimate by engineer Isaac Newton. Nonetheless, the *Monitor* did include varied and clever machinery: water closets, numerous engines, special gun mounts, skylights, and a cylindrical anchor well that allowed raising and lowering the anchor without exposing the crew. Only the turret and the pilothouse protruded above the raft, the former containing two eleven-inch Dahlgren guns and their crews, and the latter a pilot and the captain for steering and command. To aim the guns, a small steam engine rotated the turret. A major bulkhead amidships supported the turret through a central spindle. The only significant target presented to the enemy, the twenty-one-foot-diameter turret was protected by eight inches of armor plate (eleven inches in front by the gunports) and had a rounded shape to deflect any shot that should hit. When in proper alignment, four hatches in the turret floor opened to the crew spaces, which were also accessible through deck hatches. Heavy iron pendulums swung down over the gun-

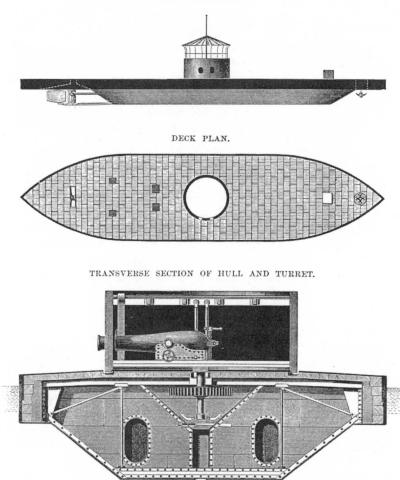

SIDE ELEVATION.

DECK PLAN.

TRANSVERSE SECTION OF HULL AND TURRET.

Three views of the *Monitor,* showing the turret, the pilothouse, and the two-hull structure. Note the position of the waterline in the turret section and the connection between the upper hull, consisting of armor plate backed with wood, and the lower iron hull hanging below it. The overhangs, where the upper hull extended beyond the lower one, caused much trouble for the *Monitor.* Reprinted from John Ericsson, *Contributions to the Centennial Exhibition* (1876; reprint, Stockholm: Royal Swedish Academy of Engineering Sciences, 1976), 22.

ports in the turret to protect the crew during loading. This round, steam-powered, rotating turret became the *Monitor*'s most lasting contribution to naval architecture.

The unique design, which put all the crew spaces except the turret and the pilothouse below the waterline, made the *Monitor* part submarine. The difficulties of supporting life when the crew was sealed belowdecks were only one step removed from those of a completely submerged vessel. The decks were awash in all but the calmest sea, so the entire *Monitor* sealed shut with the exception of air vents. This structure also cut the crew off from awareness of the outside world. Only the captain and the pilot, peering out from slits in the armor, could see out when the vessel was trimmed for battle. Heavy pendulums covered the gunports, the only openings in the armor; thus, the gun crew, though they breathed outside air, could peek out only when the guns retracted from the ports.[35] The internal spaces of the *Monitor* constituted a mechanical world: machines supported life. As designed, this world defined the crew's experience, protecting them from the dangers of the surrounding ocean and the enemy.

The ambitious construction project got quickly under way. Within a few weeks of signing the contract, the group hired Thomas F. Rowland, who owned Continental Ironworks of New York, to build the hull. To manage the project, Ericsson gained control over Rowland's workforce and constructed a special building in his yard to house the building efforts.[36] Soon he had 175 men on the project, and they began laying deck beams. Rowland reported, "They are making quite a show." Ericsson personally directed the relationships between the contractors; his acquaintance with the owners was key to expediting the project. Rowland reported some delay in receiving the iron for construction, but he said,"The Troy folks have been much more prompt than rolling mills generally are."[37] Ericsson produced drawings and specifications from his office at a furious pace (although a complete set for the *Monitor* never existed). He regularly visited the contractors to personally examine the work.[38]

Ironclad vessels proved an ideal way to bring unused industrial capacity to bear on the war effort. During the fall of 1861 and the winter of 1862, a variety of firms in the area surrounding Greenpoint, New York, built the *Monitor*. Pieces were parceled out to subcontractors—nearly all clustered in New York City and Albany, centers of the steam-engine and iron industries. The *Monitor*'s industrial pedigree resembled that of a floating locomotive as much as a warship. Several of the companies normally built railroad parts and machinery but had fallen on hard times during the 1850s. Delamater built the boilers, the main engines, the propeller, and other machinery. Clute & Brothers of Schenectady built the turret engines, the gun carriages, and other small machinery. Novelty Ironworks of New York fabricated the turret (Novelty had a close relationship with Engineer-in-Chief Benjamin Isherwood and was also building the navy's "ninety day gunboats," under his direction). Abbott & Sons of Baltimore rolled the iron plate. Rowland's Continental Ironworks built the hull. In addition to building future monitors, several of these firms later played important roles in American heavy industry. Novelty built Isherwood's monumental fast cruiser *Wampanoag* in 1863. Winslow, Corning, and Griswold, along with Alexander Holley, set up the first American Bessemer steel plant at Troy in 1865. In the 1870s, Delamater built elevated railroads in New York City.[39] As Ericsson hoped, the *Monitor* represented the strengths of American industry. The press covered the enterprise with great interest, although reports that reporters considered the project "Ericsson's folly" are exaggerated.[40]

Commodore Joseph Smith, a member of the Ironclad Board, was Ericsson's prime contact with the navy during the construction period. The two had had at least some contact before, because Smith had installed an Ericsson caloric engine in the ordnance building at the Washington Navy Yard.[41] Nevertheless, Smith found the duty a burden and constantly worried that the ironclads would prove a failure and hurt his reputation. The correspondence between the two men discussed details of armament, iron plating, and basic dimensions. Throughout, Smith retained his original skepticism, forcing Ericsson to anxiously confirm his confidence. Ericsson had to argue that what he presented to the Ironclad Board was "never intended as a working drawing," but rather "a diagrammatic and theoretical exposition of the *battery*," which now required modification.[42] Whereas Smith expressed the anxieties of the sailor, Ericsson countered with the numbers of the engineer. "There is no living man," he wrote to the commodore, "who has tripped me in calculation or proved my figures wrong in a single instance in matters relating to theoretical computation." Even so, the sheer extent and speed of the project overwhelmed Ericsson. "The magnitude of the work I have to do exceeds anything that I have ever before undertaken because there is not sufficient time left for planning, everything must be put in hand at once, a condition truly difficult. I beg you to rest tranquil as to the result; success cannot fail to crown the undertaking. Nothing is attempted not already well tried, or of so strictly mechanical a nature as to be susceptible of previous determination."[43] This final statement crystallizes Ericsson's philosophy: "previous determination," that is, drawings and calculations, formed the core of the problem; execution and construction were "strictly mechanical." Ericsson's neglect of the pragmatic details required to embody his designs proved to be the *Monitor*'s downfall.

Despite the assurances, early in the construction process Smith began to question the unique human arrangements in the new vessel. He worried that people would not fit properly into the design. "How is a man of five feet eight or ten inches in height to stand at work in a conical iron pilot house only five feet high or less if the dimensions are outside measurements?"[44] How would the sealed vessel, he asked, vent human waste when under fire? Smith warned about the morale problems of keeping sailors below in the iron hull: "Your plan of ventilation appears plausible but sailors do not fancy living under water without breathing in sunshine occasionally."[45] Ericsson responded to each of Smith's entreaties with supercilious confidence, but eventually he tired of the continuing questions that distracted him from the intense, exhausting drawing. "I intended to have sent an elaborate reply to your remarks relative to the strength of the battery, but abstain because at present I am hard pressed for time, and it is an unpleasant task continually to contradict the opinions you express."[46] Much to Ericsson's dismay, from its earliest stages he had to defend his design.

Smith's opinion was not the only uncertainty surrounding the *Monitor*'s construction. The contract dictated that the vessel be delivered in one hundred days, an impossibly short time. Furthermore, the combat clause in the contract stipulated that even though the vessel was a prototype, it was not to be fully accepted by the navy until proven under enemy fire, and the builders were liable for the entire cost until that time. Just what success or failure would mean in these circumstances, however, the contract did not specify. A victory in combat would certainly make a strong case. Thus, Ericsson and his partners had a financial incentive to rush the new ship into battle as quickly as possible; a public showing would provide evidence of the successful fulfillment of their contract. Furthermore, two other projects, the *New Ironsides* and the *Galena,* were also under construction, competing with the *Monitor* to be the core of a true ironclad fleet (the men who invested in the *Galena* were the same as those who invested in the *Monitor,* so success for either would bring them rewards).

Ericsson and his associates competed not only with these other private projects but also with other programs arising within the armed services. The army was constructing a series of iron-casemated boats designed for fighting on the western rivers. Known as "Pook's turtles" or the "city class" ironclads (because they were named after port cities such as *Cairo, Cincinnati,* and *Pittsburgh*), these vessels resembled western steamboats with their light-draft and paddle-wheel propulsion. They eventually were operated by the navy and turned in solid, although flawed, performance during the war, including at Forts Henry and Donelson. They never generated the same euphoria, however, that the *Monitor* incited in the East.[47]

Another, more direct challenge to Ericsson's plan arose from within the navy, indeed from its own expert officers. In the fall of 1861, Engineer-in-Chief of the Navy Benjamin Isherwood developed a plan for an iron-plated steam battery, to be built in less than six months for about one-half million dollars. His partner and mentor, John Lenthall, a highly regarded naval constructor, had recently been appointed chief of the navy's Bureau of Construction and

Repair. Lenthall had a solid reputation for naval engineering behind him, having designed numerous ships, including the original *Merrimack*. Like the *Monitor*, the new vessel proposed by Lenthall and Isherwood employed the cylindrical turret (two turrets, in fact). Unlike Ericsson's turret, which mounted on a central spindle, this one, based on the design of English engineer Cowper Coles (much like Ericsson, an independent at odds with the navy), rested on bearings around its circumference. Ericsson considered the proposal an affront; he always believed that Coles had stolen the idea from him, even while insisting on its fundamental difference from his own. Nonetheless, the Lenthall-Isherwood threat came from within the naval establishment and had to be taken seriously. At issue were not only the details of design but also who would design warships: civilian inventors or military experts.

This abstract question became a matter of practical politics. A bill to finance the Isherwood-Lenthall design for twenty batteries made its way through Congress, and Ericsson's allies acted to suppress it. Congressman Erastus Corning wrote to Gideon Welles that he was "apprehensive that the plans adopted or proposed for adoption, are not such as will secure the Government the very best result . . . hasty and immature action should not be had in determining this question."[48] John Griswold went to Washington and pressured Assistant Secretary of the Navy Gustavus Fox.

From this point on, Fox played a central role in the *Monitor*'s history. An experienced naval officer (he had even served on the *Princeton* in 1853), Fox had retired in 1856 to the manufacturing business in Lawrence, Massachusetts. In 1861 he returned to the navy to assume the newly created position of assistant secretary of the navy. Working closely with Gideon Welles, Fox managed the operations of the navy department during the entire war, gradually becoming a powerful proponent of Ericsson's cause. He witnessed the Battle of Hampton Roads, and as the first person aboard afterward, he instantly declared the *Monitor*'s victory "one of the greatest naval conflicts on record." When Griswold first went to see Fox, however, the assistant secretary was not yet an ally. Griswold reported, "I was determined he should either convert me to the bureau [Isherwood] plan, or I would him to our plan." Griswold pressed him on the matter of professional expertise. "I left him with an admission that he was only familiar with *sailing* and *defending* a ship; that, as to the mechanics and architecture of incident to a ship or steamer building he professed to know but little, and so far as the mechanical and other arrangements of the Ericsson battery were concerned, he would concede to me that it appeared to embody all the features of success." Fox granted that if the *Monitor* enjoyed a successful demonstration, the bureau would adopt the plan of Ericsson and his allies. "I made a convert of him," the satisfied Griswold reported, "and felt abundantly compensated."[49]

The efforts in Washington gradually framed a potential outcome for the "experiment" of the *Monitor*. "I assure you," Winslow wrote to Ericsson, "the energy and despatch exhibited in the construction of this battery is unequalled . . . and will give us a position and influence with the Government in any future contracts that will be almost controlling . . . if the battery comes up to what we have promised, I tell you in all sincerity that other plans and other contractors *will be nowhere*. Our 'prestige' will be hard for others to overcome."[50] Griswold echoed these sentiments and for the first time introduced the notion of a spectacle to convince the remaining doubters: "I cannot imagine that the 'twenty gun boat' conspirators [Isherwood & Lenthall] have grounds for hope—certainly not if the 'Monitor' is demonstrated first."[51] Griswold soon declared the question closed. "At least the Navy Dept. will not authorize more than one or two boats on the Isherwood plan until ours is put in proof, and if that proof is satisfactory, I have a promise from the *very* highest source that we shall have all we want of the twenty to be built. This *ought* to satisfy us."[52]

His mission completed, Griswold left Washington and returned to Troy, but he retained his lawyer to act as a lobbyist or a "proxy" to track further progress and make sure things did not unravel. This man, M. L. Barnes, made the rounds of the navy department, visiting Welles and Smith. He reported to Ericsson in early February that the "Gun Boat Bill" had passed the Senate but that the navy leadership favored building more monitors. "I think your efforts will be highly appreciated . . . we will bring strong influence to bear here," he said, and again the following week, "Your plans are in favor . . . There is a general expectation of important results from her actions . . . I almost incline to congratulate you in advance of a complete experiment."[53] Gradually, it became clear that the navy was willing to suspend its new contracts pending a successful demonstration of the new vessel. Bushnell, in Washington in late February, reported from the Willard Hotel that once the vessel had been shown successful in battle, "diplomacy will be unnecessary to secure all that you desire." Bushnell reemphasized that everything depended on a successful demonstration: "No plans drawings or anything of the kind have been made yet for the proposed 20 ironclad vessels—in fact I have it from the highest authority that everything depends upon the test of your battery and that until after her trial nothing will be done."[54] Winslow, again in Washington in early March, reported that the appropriations bill had passed but that it left the matter of choosing the vessels' type and contractors to the judgment of Secretary Welles. It was critical, then, to convince him of the *Monitor*'s success.[55] Without Ericsson's personal involvement, the *Monitor* builders had all but eliminated a significant, and highly qualified source of competition. The victory, however, was only temporary; it needed to be sealed by a convincing demonstration.

Despite the progress in politics, the *Monitor* construction lagged. At the end of the one-hundred-day contract period in mid-January, Smith wrote ominously to Ericsson, "The time for the completion of the Shot-proof Battery according to the specifications of your contract, expired on the 12th instant," and more directly on February 3, "She is much wanted *now*."[56] Toward the end of February Fox grew nervous as well; he had reports that the *Virginia* would soon head toward Hampton Roads. Fox telegraphed Ericsson, "It is very important that you should say exactly the day the *Monitor* can be at Hampton Roads."[57] Griswold, too, expressed his concern to Ericsson, but for reasons as much contractual as patriotic: "I trust the *Monitor* may arrive safely at Hampton Roads in time for the *Merrimack* . . . I shall feel extremely anxious for the fate of the battery . . . Our requests for remittances then can receive immediate attention."[58] Though they had headed off the Isherwood and Lenthall plan, the *Monitor* builders still faced competition from the other two original ironclad contracts, the *Galena* and the *New Ironsides*. Public demonstration would enable the *Monitor* builders to solidify their advantage over all competitors. For these men, then, technological display of their invention was a material requirement for the success of their business enterprise. The encounter with the *Virginia* was virtually written into the *Monitor*'s contract.

Fitting Out, Populating

The *Monitor* was launched from its berth at Rowland's Continental Iron Works on January 30, 1862, and spent several more weeks there fitting out. At the end of February it was put into commission and received its crew, comprised primarily of volunteers whom Captain John Worden had selected from the *North Carolina* or the *Sabine,* two ships in the New York Navy Yard. Worden, 43 years old but still a lieutenant, reflected the difficulties of promotion in the antebellum navy. This was his first command. In April 1861 he had been intercepted while delivering a message in Alabama and had become the first prisoner of war in the South. By the time he was released in mid-November, the *Monitor* was already taking shape. He reported for duty January 16, before the vessel was complete.[59] In contrast, naval academy graduate Lieutenant Samuel Dana Greene, the vessel's executive officer, second in command, was merely 20 years old. Greene's performance in the battle of Hampton Roads was a primary topic of debate for the crew, years after the war. Isaac Newton, chief engineer on the *Monitor,* was an experienced steamboat captain and engineer and became one of John Ericsson's few friends later in life. Second Assistant Engineer Albert C. Campbell actually operated the machinery under Newton's command. Other officers aboard the *Monitor* included John Webber and Louis Stodder, acting masters; Daniel S. Logue, acting assistant surgeon; and R. W. Hands and M. T. Sundstrum, third assistant engineers. Acting Assistant Paymaster William Frederick Keeler joined the ship January 12 and oversaw all accounts, paybooks, and provisions.

Fireman John M. White recalled that the enlisted men who volunteered for the *Monitor* "were made all manner of fun by those on bard for gooing to sea in a tank as most of them term it." He shipped on the *Monitor* under an assumed name, John Driscoll (a not-uncommon practice at the time).[60] Most of the forty-eight men were three-year volunteers in their twenties and lived near New York City (the oldest was thirty-eight and the youngest eighteen). The mix included men born in Ireland, Sweden, Norway, and Wales, typical of a nineteenth-century muster roll. Two were black and at least one Jewish. On the muster roll, most indicated "none" under occupation, but a few listed carpenter, stone cutter, chandler, or machinist. One who listed "sail maker" undoubtedly had to find other things to do.[61] The *Monitor's* account books do not survive, but those of the *Passaic* (Ericsson's second monitor) reveal that on that ship all but three of the seventy-seven enlisted men signed for their pay with an "X," suggesting a low rate of literacy aboard the *Monitor* as well.[62] Hence, the fifty or so enlisted men on the *Monitor* left fewer written documents than the officers did. Even so, some accounts of the battle and the sinking survive, written by the men years after the battle. Fireman George Geer joined the *Monitor* thanks to a hometown connection with John Griswold; he left a complete set of letters to his wife from 1862, presenting the most complete seaman's view of life aboard the ironclad.[63]

Although not technically part of the crew, one other engineer became an important player in the *Monitor's* history. During construction, Ericsson believed "an engineer of the highest intelligence will be required" to inspect the vessel and act as overseeing agent for the government, "success mainly depending on that officer as the whole is machinery." Ericsson suggested to Joseph Smith that he appoint Chief Engineer Alban B. Stimers, and Smith concurred. Stimers, 34 years old, had joined the navy in 1849 and had become chief engineer in 1858. He had served as chief engineer aboard the *Merrimack* and had recently been on a board evaluating the completion of the long-overdue Stevens Battery. Beginning in November, Stimers superintended construction of the *Monitor* and learned its machinery in detail.[64] When the vessel entered commission, Ericsson thought Stimers the man most familiar with its operation, so Stimers accompanied the *Monitor* to Hampton Roads as an observer. Though not officially assigned to the vessel, Stimers performed many of the chief engineer's duties, repairing and operating the machinery until Isaac Newton and his engineers could grow fully acquainted with it. Aboard the *Monitor,* Stimers was Ericsson's man, reporting praise and technical details to him in New York. In the naval hierarchy, Stimers actually worked for Ericsson's rival Benjamin Isherwood. He had no use for his chief's low opinion of the *Monitor* design, but Stimers did agree with Isherwood that engineers did not receive their proper status in the new industrial navy.[65]

Stimers had ambitions beyond Ericsson's tutelage. Before the ship was completed, the young engineer struck up his own correspondence with Assistant Secretary Fox, laying the groundwork for a program of his own. "I have observed the difficulties [in construction] which have been avoided and those that have been encountered," he wrote to Fox in early February. Voicing nothing but respect for the master Ericsson, whose methods Stimers could deeply appreciate, he nonetheless also wrote of the problems caused by such rapid construction: "In the case of the *Monitor* there is very great want of exactness and detail in the specifications, which makes the duties of the superintendent very responsible and onerous." Stimers eagerly desired "the privilege of going out in the *Monitor*" but hoped that "when she has proven her capabilities," he could be spared to help specify future generations of monitors.[66] Stimers stood at the nexus of conflict between public and private engineers, a clash already stirred by Ericsson's campaign against the Isherwood-Lenthall proposal. Zealous Stimers had much to gain from his association with Ericsson, but also much to gain from his insider status in the navy. By participating in the construction and operation of the *Monitor* at sea, Stimers hoped to mix the expertise of the engineer with the experience of the naval officer. This notion sowed the seeds of deep trouble, for the "light draught monitors" that Stimers built later in the war turned out to be a scandalous failure.

Captain John Worden put the ship in commission February 25. During the following week, the vessel embarked on a series of sea trials. The first day out revealed a number of problems with the steering gear and with the direction of the propeller, which Ericsson easily repaired. A few days later, on another trial, the guns were fired several times. Although the mechanisms were damaged (for which Ericsson blamed Stimers), the ship was deemed ready for combat. On March 6, 1862, the *Monitor* left New York under orders to head for Hampton Roads.[67]

During construction, while it was still under his potent care, the *Monitor*'s artificer could control what his creation would communicate. Ericsson's messages and those of the machine were one and the same. He determined the shape, the architecture, even the name. An array of powerful allies supported the invention. Persuaded by the engineer's authority that the idea would succeed, they silenced opposition with political pressure. Now that the vessel was completed, however, the *Monitor* builder had to send it out into the world to be run by strangers and to meet hostility, both natural and human. The inventor's language and that of the machine would diverge, and no amount of effort could reunite them. Ericsson hoped the ship would speak "success," "victory," and "revolution," but once separated from its maker, the *Monitor* belonged to its users and to its observers. People like William F. Keeler would have messages of their own.

Chapter Three

William Keeler's
Epistolary *Monitor*

PUBLISHING POETRY IN A MAN-OF-WAR

By advice of a friend, Lemsford, alarmed for the fate of his box of poetry, had latterly made use of a particular gun on the main-deck, in the tube of which he thrust his manuscripts, by simply crawling partly out of the port-hole, removing the tompion, inserting his papers, tightly rolled, and making all snug again . . .

Breakfast over, Lemsford and I were reclining in the main-top . . . when, of a sudden, we heard a cannonading. It was our own ship.

"Ah!" said a top-man, "returning the shore salute they gave us yesterday."

"O Lord!" cried Lemsford, "my Songs of the Sirens!*" and he ran down the rigging to the batteries; but just as he touched the gun-deck, gun No. 20—his literary strong box—went off with a terrific report.*

"Well, my after-guard Virgil," said Jack Chase to him, as he slowly returned up the rigging, "did you get it? You need not answer; I see you were too late. But never mind, my boy; no printer could do the business for you better. That's the way to publish . . . fire it right into 'em; every canto a twenty-four-pound shot."

Herman Melville, *White Jacket.*

My hands are all dirt & powder smoke as you will discover by the paper.

William F. Keeler to Anna Keeler, March 9, 1862,
immediately after the Battle of Hampton Roads

The significance of a technology, especially in war, is in large part created by the stories people tell about it. Melville's anecdote linking writing and cannonballs drives the point home. Whereas Ericsson fired drawings from his pen, William F. Keeler sent letters to his wife. He tells a story of the *Monitor* at once darker, more intimate, and less sure than the canonical, revolutionary narratives. Before we examine the *Monitor*'s operational life, then, it is worth considering the conditions under which Keeler wrote and how they framed his views of mechanical warfare. The vantage point of experience did not always match that of Ericsson's engineering.

Who Was William Frederick Keeler?

As the *Monitor* itself did, William Keeler's naval service juxtaposed appearance and experience. After a three-day train trip from his home in Illinois, he arrived at the New York Navy Yard on January 12, 1862—the hundredth day of the *Monitor*'s construction, the day the unfinished vessel was due for delivery. "We don't get a great many sailors from the prairies out there," remarked the officer who greeted Keeler. Most of the men who lived so far from shore chose the army instead of service at sea. He thought he would not need a uniform. When asked if he had one, the landlocked Keeler thought "perhaps they would let me go as I was." "Not at all, get a uniform before you go to sea," came the blunt reply.[1]

Keeler bought a uniform, giving sartorial expression to his induction into the navy. The new clothing, like the new environment, felt uncomfortable at first. "Bright handsome uniforms are so common here that scarcely any notice is taken of them," he wrote to his wife, Anna. "I felt awkward enough at first in mine, it seemed as if every one was looking at me, but I am getting used to it now."[2] Keeler looked like an officer, but the change inside took longer.

Soon after his arrival, Keeler visited a photographer and sent Anna a picture of himself in his new uniform. Both thought the portrait stiff and unnatural. As Keeler posed, he had to follow the hurried photographer's wishes, because other men stood in line to have images of their own naval selves to send home. "I wanted to select my own attitude and put on spectacle frames, but the artist said it was unnecessary, I am sorry now that I did not insist upon having my own way."[3]

Keeler's carriage did not yet fit a standardized naval demeanor, but gradually he learned the benefits of belonging. While still in port, he found himself mixing in polite society. "I find that they (buttons &c) are a sure passport to the notice of the weaker sex & I rather enjoy the idea of handing those around who, if I was dressed in other clothes, would scarcely notice me." For Keeler, the emblem of his new experience was a newfound attention from women:

> Our vessel has been visited by hundreds of ladies who heartily expressed good wishes we shall carry with us. The duty devolved upon me to shew most of

them around the vessel. You can imagine your polished & accomplished husband *shining* in this new sphere—I believe I got along well enough. I rubbed my antiquated & somewhat indistinct ideas of etiquette & bright buttons and Shoulder straps made up any deficiency.[4]

Soon after, Keeler attended the launching of a new steamer, the *Adirondack*. The Brooklyn Navy Yard came alive with visitors and pomp on a Saturday afternoon. Warships, decorated with flags and streamers, fired salutes. Brass bands played "our favorite national airs." Amid the celebration, Keeler's uniform, and its shiny buttons, admitted him to the day's high honor: "The greatest treat to me was the launch. I was one of the privileged ones admitted on board (buttons again) & it was an exciting scene as she [the *Adirondack*] slid gently off the ways into her future home amid the roar of cannon, the music of the band, the waving of flags & the cheers of thousands."[5] What a thrilling introduction to naval life for this fresh officer from Illinois! Just weeks before, Keeler had thought he could do without a uniform. Now he found himself amid the naval elite, with all the attention and prestige it entailed. Keeler frequently reminded his wife that accompanying these powerful men were beautiful wives and daughters.

In the year that followed, Keeler went from a novice volunteer to an experienced and sometimes jaded naval officer, shaped by the adventure, horror, and tedium of war. His long, detailed letters to his wife, Anna Elizabeth Keeler, chronicle this transformation. Keeler provides a singularly rich account of the *Monitor*'s story, not as a detached witness, but as a participant who experienced the full weight of the circumstances he describes. His letters are interesting not only as a source of information (although "being there" is no objective position, introducing its own distortions) but, with their amalgam of motive, tone, and intention, also as an extended meditation on the personal experience of technological change in the nineteenth-century navy.

William Frederick Keeler was born in New York in 1821. He lived in Bridgeport, Connecticut, for a time and in 1846 married Anna Dutton, the daughter of a prominent Connecticut lawyer, Henry Dutton, head of the Yale Law School and later governor of Connecticut (perhaps Dutton's connections helped place Keeler on the *Monitor;* Cornelius Bushnell also came from New Haven). The Keelers settled in LaSalle, Illinois. There William opened a watch and jewelry business in 1853, which was "doing a good business" by 1855.[6] In 1857 he sold the business and became the senior partner of Keeler, Bennigin, and Company, which advertised itself as "La Salle Iron Works, Founders & Machinists, on the Steam Boat Basin, Manufacturers of Steam Engines, Mill Gearing, Horse Powers, Corn Shellers, Coal Cars, Stoves, Iron Railing & Machinery of all kinds. Brass & Iron Casting of any description made to order."[7] Keeler probably had a flair for machinery. An avid reader of the *Scientific*

American, he enjoined Anna to maintain his subscription during his naval service. He had an early thrill when he showed an editor from the magazine around the *Monitor.*[8] Keeler's partner left in 1859, "worth nothing," and the business was rated "doubtful" for credit by R. G. Dun and Company.[9]

Though living in a landlocked state, Keeler had had two years of sea experience, having sailed to California around Cape Horn during the gold rush of 1849, on to China, and then back to New York (a trip on which two of his brothers died). During the war he ran across at least one of his former shipmates.[10] Keeler condemned slavery as a "hideous deformity" and believed the Confederate enemy to be composed of traitorous, wretched souls. He once observed of a church in Jamestown that "joining it was a deserted rebel earth work, bringing in strong contrast the deep toned piety of the early settlers & the perjured villainy of their degenerate offspring."[11] A great supporter of General McClellan (whose name he regularly misspelled), Keeler frequently criticized the naval command but never questioned the motives or the justice of the war.

Keeler joined the *Monitor* as acting assistant paymaster, a new position created by the navy to support the clerical work in its rapidly expanding force. He kept the ship's accounts—a position requiring a great deal of writing and a certain degree of arithmetic skill, which Keeler no doubt brought from his experience of running a business.[12] It appears that Keeler underwent no training upon his induction into the navy. He ordered and parceled out the ship's provisions, including clothing, many pounds of candles to light the wardrooms, beeswax, ribbon, tape, shaving supplies, and food. He probably earned about fifty dollars per month, an amount equal to that of the engineers and second only to the pay of the captain, who made three times as much. The seamen made between four and fifteen dollars per month.[13]

At 40, Keeler was the oldest officer on the ship except for the captain. He tended not to mix with "the men" and spent much of his time with the ship's surgeon, who had the poor taste to read long, dull, love letters to the business-like paymaster. Teetotaler Keeler railed against drink and hence frequently disapproved of the crew's behavior. "There are three great evils in both our army & navy which if corrected would render them much more efficient— the first is whiskey, the second is whiskey, & the third is whiskey."[14] Keeler's position as paymaster increased his social distance from the crew, although the captain made the real financial decisions. When the ship ran out of money or the captain withheld pay, the crew blamed Keeler, sometimes questioning his competence. George Geer once referred to him as a "devilish scoundrel."[15] One episode hints at the enlisted crew's attitude toward their older, dry clerk: "Last night an empty wine bottle came sailing over the top of the partition into my room about 12 o'clock & its crash echoed all over the ship . . . Everyone looked very innocent at the breakfast table this morning. The youngsters

William Frederick Keeler, soon after receiving his uniform, January 1862. "I wanted to select my own attitude," he wrote to his wife, Anna, of posing for this photograph, "but the artist said it was unnecessary, I am sorry now that I did not insist upon having my own way." Courtesy of Naval Academy Museum, Annapolis, Md.

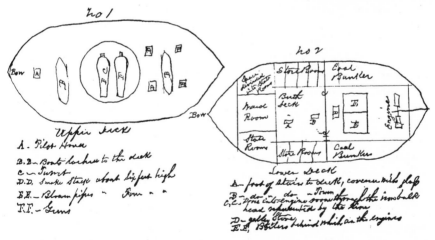

Keeler's sketch of the upper (1) and lower (2) decks of the Monitor. Officers slept in the staterooms; crewmen slept in hammocks on the berth deck. From a letter from Keeler to his wife, March 7, 1862. Courtesy of Naval Academy Museum, Annapolis, Md.

are full of tricks."[16] Amid the crew, Keeler had that tinge of social awkwardness that makes a good observer. But his background and his literacy made him better prepared than the other officers to mix with the ship's elite visitors.

Keeler's Epistolary *Monitor*

It is not a coincidence that the paymaster, who was neither a technical expert nor a career navy man, neither immersed in technology nor vested in the naval tradition, produced the most thorough account of life aboard the *Monitor*. Those operating and maintaining the vessel and its machinery rarely had time to reflect. Keeler is a fine example of a participant observer. He spent long hours in his room, composing descriptions of the ship, the life aboard, and his own impressions. He wrote seventy-nine letters to Anna over a twelve-month period, from January 1862 until the following New Year's Eve, when the vessel sank. The letters themselves are artifacts, preserving the physical effects of the days' events. "My hands are all dirt & powder smoke as you will discover by the paper," he wrote after the battle in his March 9 letter. Keeler's fingerprint appears, clearly etched in black powder, near the date in the top right-hand corner.

Sometimes he included sketches of his room, of the vessel, of the disposition of warships along the James River. The officers' staterooms surrounded the wardroom, the common area where they would relax and dine. When the *Monitor* sat idle on blockade, Keeler wrote long, mundanely detailed letters to while away the time. He included a range of information, from war stories in the wardroom to the details of Christmas dinner. He once sent Anna a table of rations to show what he ate.[17] The mood of his letters similarly reflected the situation aboard ship. During busy and exciting times, they were short and businesslike, becoming more detailed only when time could be found to narrate the events. When the insides of the *Monitor* grew hot, Keeler, like the enlisted men, sweated profusely as he wrote.[18]

Keeler's correspondence displays a candor common to Civil War letters.[19] Military secrecy did not restrain the narration. He served on the most advanced weapon in the fleet, yet he brought his father-in-law aboard to tour the ship before it was completed, showed it to a reporter from the *Scientific American* (which published a description), and invited personal friends aboard for tours.[20] Keeler freely described the *Monitor* and its activities to his wife, who showed the letters to friends and even had them published in the newspaper. He detailed the vessel's dimensions, its operations, its command structure, its strengths and weaknesses, even the debates surrounding its use. Keeler gathered this information from the highest levels. "Sec'ys of the War & Navy, Generals, Cols., Commodores & all intermediate ranks & grades assemble in our cabins to discuss the pros & cons & every word is distinctly heard in our

State rooms, so that we become possessed of knowledge which would send some of our editors to Fort Lafayette [prison] if they were so indiscreet as to publish it."[21] George Geer divulged similar details to his wife, merely asking her to be careful not to speak of them to any secessionists.[22]

The paymaster did have a notion of military secrets, but the regulations depended more on trust than on coercion for enforcement. "We rec'd an order from the Department today forbidding any one on board giving information to any one in regard to the vessel, her armament, repairs, intended destination, or any of the details of construction & no one must be allowed on board except the personal friends of the officers, in whom they have the fullest confidence—this applies to all the vessels."[23] When Anna asked about the *Monitor*'s movements, Keeler reminded her that "not a newspaper reporter has been allowed to come up the river since it has been opened & orders have been sent to all on board the vessels to write nothing for publication that would give any information of our movements."[24] Curiously, the *Monitor* became secret only after its much-heralded public performance. Nevertheless, Keeler apparently did not include his own words on the list of dangerous speech and generally gave Anna details of plans, destinations, and fears.

In the busy week before departing, Keeler had written only briefly to Anna, saying, "When we are once at sea & everything is reduced to a system I will give you all the details of this new sphere of existence." With the vessel under way, Keeler sat down to write more: "Perhaps you would like to know just how my room looks—I wish you could look into it & see it for yourself . . . all of nice white ware with "Monitor" on each in gilt letters. Tapestry rug on floor, The berth, drawers, & closets are all of black walnut, the curtains are lace & damask, or an imitation I suppose . . . [I] have seen no room as handsomely fitted up as ours."[25] Curiously stylish for an industrial tool of war, this comfortable chamber was the stage for Keeler's letter writing.

During the day a skylight allowed the sun in; because the deck was frequently awash, Keeler sometimes composed his letters in daylight filtered by the ocean. Usually he wrote in the evening, his room darkly lit by a candle, often filling every space on the page, even if it required fitting quips sideways in the margins. It pained him to send off an incomplete letter, but naval operations frequently required brevity, "I am full of write, but unfortunately time is wanting" he complained.[26] Walls divided the rooms in the *Monitor* but did not reach to the ceiling, so Keeler overheard the officers in the wardroom "as if they were seated by my elbow. This does not assist one to concentrate his thoughts & very likely is the occasion of not a very few mistakes."[27] Occasionally he made a revealing slip, the language of destruction never far from his pen: "The events of a lifetime seem crowded into the last few days & I can hardly convince myself at times that the past week has not been an exciting Shell (they are talking about shell in the Cabin) dream."[28]

As war letters often do, Keeler's words oscillated between the tragic and the mundane: "I rec'd the "Press" this morning—was sorry to see so many of my acquaintances among the killed—such things bring the war home and we feel it to be a sad reality—I hope your spell of weather will be over before long but as long as it does continue, take my advice & don't put linen sheets unless you make a good coal fire on top."[29] Keeler saved the letters he received from Anna—against advice, lest they be captured (seaman George Geer burned his after reading them). Unfortunately they do not survive (shortly before Keeler's death, Anna wrote several letters for her ailing husband to a *Monitor* enthusiast in New York, and these are the only ones in her hand that survive), nor do those he wrote to his children.[30] Keeler did send his son Henry a shiny button from his uniform, and he sent the family a shell fragment from the *Virginia.*

Chronicle and detail notwithstanding, Paymaster Keeler had more than time. He also had a literary flair. His letters offer a series of reflections on war's human and mechanical faces. Consider his descriptions of colleagues. The ornery first lieutenant Samuel Dana Greene had "black hair & eyes that look through a person." The captain, John Worden (manliness never being far from Keeler's mind), "is tall, thin, and quite effeminate looking, notwithstanding a long beard hanging down his breast—he is white & delicate probably from long confinement & never was a lady the possessor of a smaller or more delicate hand, but if I am not very much mistaken he will not hesitate to submit our iron sides to as severe a test as the most warlike could desire."[31] Like many soldiers, Keeler was struck by the sublimity of war's spectacle, as he wrote of the fireworks at Malvern Hill:

> Of course nothing could be seen in the darkness but two long parallel lines of black wall between which wound the narrow stream we were navigating. The dancing light of myriads of fire flies sparkled in strong relief against the dark back ground, startled from their leafy covert by the heavy concussion of our guns.
>
> The scene was one of the most terrible magnificence & will remain forever impressed upon my memory, it was a sight I have often desired to see, a bombardment in the night.[32]

Keeler's Readers

The mix of personal and profound, reporting and ruminating, in Keeler's letters raises a question: for whom were they written? The obvious answer, his wife, Anna, and their children, is clearly and exclusively correct at the beginning. Soon, however, Keeler realized he was living through unique circumstances, even in a war full of compelling individual narratives. Immediately after the battle of Hampton Roads, he wrote a long letter describing the events. For the first time, he conveyed a sense that Anna and his children might not be the only readers. He excuses his rushed prose, "If [the letter is] read to any out of the family you must apologize." And again the following week, "My letters must be a confused, jumbled up, disconnected mass of almost nonsense. I hope you polish them up before reading them to others." Soon it became necessary to distinguish private letters from others, "I think you will find this a particularly stupid letter & had better consider it private for that reason."[33] He began numbering his letters to be sure Anna would know their place in the historical narrative, regardless of the vicissitudes of the wartime mails.

Before long, Anna began distributing her husband's letters. She sent his account of the battle to her father, Henry Dutton, who sent it to the *New York Times,* where it appeared on the front page. "Interesting particulars by one on board," ran the headline, which did not name Keeler but revealed its author to be the ship's paymaster. Keeler wrote home when he found out about the article: "Of course I cannot doubt his motive—it was good, but it has caused me the greatest uneasiness." He did not recall what he had written and feared it might cause discomfort for his fellow crew members (but as it turned out, visitors to the *Monitor* complimented him on the account). He then asked Anna to send no copies of his letters to anyone without prior approval—as much out of concern for language as for politeness.[34] By this time, however, the *Monitor* had become famous, and Keeler a minor celebrity in his own town. A week later he wrote, "I do not care about your reading my letters to our friends when you think them suitable, but I do object most decidedly to your lending them . . . I want all to understand that my letters are not written with the expectation of having them criticized—they are merely familiar scratches to the dear ones at home, written as I have said before amid confusion enough to distract a ship load of mules."[35] When the editor of the *LaSalle Democrat-Press* became interested in publishing Keeler's observations, Keeler pleaded with Anna for discretion. His concern centered chiefly on style: "Don't let any loose, badly expressed sentences get into print. The editor should furnish each of us with a copy as the least thing he can do. Keep my reputation in sight & use your own judgment in the matter."[36] His interest in publication went beyond news, for the self-conscious Keeler desired his friends to know of his exploits, to enhance his social standing at home. "I expected to have rec'd letters from all my Lasalle friends (male & female) congratulating this distinguished fellow citizen of Monitor renown &c &c but begin to fear that I shall be disappointed. Remember me to all who inquire."[37] In the end, Keeler did not allow the hometown paper to publish his writings, probably because of the growing criticisms of the *Monitor's* failure to pursue the *Virginia* after the battle.[38]

Anna's handling of Keeler's letters, distributing them to family and friends and negotiating for their publication, emphasizes the social aspect of his war experience; it also marks Anna's critical role in it. She was no passive recipient of news. Rather, she actively mediated her husband's interactions with his community and the press and hence crafted his public image. The *Monitor* at sea, though temporarily self-sufficient, was by no means socially isolated. Especially during wartime, navy ships' crews included short-timers who, unlike professional seafarers accustomed to long periods at sea, retained primary social and familial worlds on shore. Men worried about how they appeared at home, to their friends, and in the papers.[39] Keeler regularly commented on the relative manliness of his fellow officers and excitedly noted his own contacts with women. He wished Anna to attract for him the notice of friends of both sexes. The *Monitor* crew endlessly discussed their own heroism among themselves. "Anyone could fight behind an impenetrable armor," wrote Keeler. "Many have fought as well behind wooden walls or none at all." Anxieties generated by the "impregnable" protection of iron plate led them to prove their manly courage.[40] Sailing and fighting on the *Monitor* was an exclusively male domain, but the ship's public image, and hence the heroism of its crew, intimately depended on women as visitors, readers, observers, and, in Anna's case, literal agents of publicity.

While chronicling the immediate reactions of a middle-class everyman to the new war machine, Keeler found the *Monitor* alternately thrilling, sublime, and stifling. His narrative emerged from multiple, changing motives and multiple human relationships—with his wife, his family, his community, and the nation. Some of the letters read as the homesick concerns of a father and husband, others as the careful chronicling of a historical event and its ramifications, still others as the personal record of a man experiencing new technology. Each perspective brought its own insights as well as distortions and bias. Nevertheless, for Keeler they formed an authoritative record of his war experience. Late in life, he referred to his letters as historical documents useful for settling debates on the *Monitor*'s legacy: "During my fine years in the naval service I kept all my experiences written up in the form of letters to my wife. These were all preserved and I have them bound in book form to refer to."[41] Like John Ericsson, William Keeler made his own history.

Chapter Four

Life in the
Artificial World

 [February 25] I sincerely hope our passage will be a short one. I do not expect a pleasant one . . . I just begin to realize now that I am leaving home for I hardly know what, but I hope that a few months at farthest will bring me back.

[March 6] 9 pm. I have just returned from the top of the turret. The moon is shining bright, the water smooth & everything seems favorable. The green lights of the gun boats are on our lee beam but a short distance off & the tug is pulling lustily at our big hawser, about 400 feet ahead. A number of sail are visible in different directions, their white sails glistening in the moon light. Not a sea has yet passed over our deck, it is as dry as when we left port.

Friday, March 7ᵗʰ. When I awoke this morning I found much more motion to the vessel & could see the green water through my deck light as the waves rolled across the deck.

William F. Keeler to Anna Keeler, 1862

Despite the desertion of two crew members the night before, early on the morning of March 6, 1862, in "clear, cold weather," according to the log, the *Monitor* left New York, headed for Hampton Roads. The vessel contained fourteen officers and forty-nine seamen. Below, the *Monitor*'s engineers busily completed repairs and last-minute details. To make greater speed than its own power would allow, the *Monitor* was put in tow to the tug *Seth Low,* accompanied by two gunboats, the *Currituck* and the *Sachem.* Years later, one crew member recalled that "not a whistle sounded to cheer us as we went out. Those we passed seemed to think it would be better to have played the funeral dirge than to give us the customary cheer."[1] Its construction completed, the *Monitor* still had much to prove.

As the odd vessel and its modest escort left the protection of the New York harbor, the crew began adjusting to their new arrangements. They lived literally inside the machine. The environment of iron and steam made human comforts—heat, air, shelter, dryness, and light—available below the surface of the ocean. This was an artificial world, like a mine, a tunnel, or a spaceship. Yet those aboard, especially the busy engineers, knew these comforts to be illusions whose maintenance depended on the integrity of the system. Nervous questions lurked, like the ship itself, just below the surface: could the structure and its mechanisms sustain themselves in the face of the pounding sea or the battering enemy? More realistically, just when *would* they break down? The ship thrust answers to these questions on the crew sooner than they expected.

Submarine Security and the Iron Home

Totality of enclosure distinguished life on board from that belowdecks on a traditional vessel, which, even in the deepest hold, surrounded humans with water on only three sides. The *Monitor's* unique design surrounded the crew on all four sides. Water above made an impression; Keeler and the others often referred it when describing the *Monitor* at sea, "Now we scoop up a huge volume of water on one side [of the deck] and as it rolls to the other with the motion of the vessel, it is met by a sea coming from the opposite direction, the accumulative weight seeming sufficient to bury us forever."[2] What cultural historian Rosalind Williams has written about the literary figure of the underground also applies to the nether spaces of the *Monitor:* "It is the combination of enclosure and verticality . . . that gives the image of an underworld its unique power as a model of a technological environment."[3] The submerged, enclosing machine confronted its inhabitants with a strange experience and simultaneously hid them from the rest of the world.

The first days at sea brought curious amusement with the new conditions aboard the *Monitor.* A group of men tried to make sense of their sudden confrontation with an iron home. At first it seemed an oddity, these lavishly appointed living quarters inside the cold iron. Keeler commented, "I have seen no room as handsomely fitted up as ours. The only objection is that they are too dark. I had all my writing to [do] by candlelight and lamps are always burning in the wardroom."[4] During the day, light entered through skylights in the deck, holes about six inches in diameter set with thick glass in an iron frame. These were often covered by six or eight inches of water when the deck was awash. But light penetrated the water, and "when the sun shines bright it is sufficiently light to read and write without difficulty." The captain of a later monitor noted with delight that waves sometimes deposited fish in the skylight above his head; they would swim in circles until washed out again.[5]

Machinery mediated not only light, but air as well. Through large blowers rising above the deck, a ventilating system drew fresh air into crew spaces for breathing and into the engine room for the boilers. This airflow kept the entire engine room, including the engineers, at slightly above normal barometric pressure. Keeler proudly described the novel system: "A register is in the floor for the purpose of admitting fresh air when we desire it, so you see we are provided against suffocation—the air is forced under the floor by the blowers in the engine room." Life in the finely appointed wardroom was calm, despite the ocean outside: "The dash of the waves as they roll over our heads is the only audible sound that reaches us from the outer world. One would hardly suppose from the quiet stillness that pervades our submarine abode that a gale was raging around us . . . Our life I assure you is getting monotonous enough."[6]

In good weather, enclosure and life support systems made the *Monitor* a comfortable, protected home, what Keeler referred to as "my little snuggery." As he considered daily activity aboard the *Monitor,* he saw the novelty of the experience and found it "singular" to hear the waves roll over his head while participating in the domesticities of military life: "Nothing would strike a stranger with more surprise after walking our cheerless, wave-washed iron deck than to go below and see our bright cheerful well-lighted cozy ward room with the officers grouped around the table reading, writing, or talking." At its best, the interior of the *Monitor* gave its inhabitants a sense of security; Keeler's serene reaction sometimes indicated confidence and faith in the ironclad: "I went to sleep last night to the swash, swash, of the waves as they rolled over my head and the same monotonous sound still continues and will be my lullaby tonight."[7] George Geer agreed, finding the *Monitor* at first more comfortable than his previous home. "We lie much better here than we did on that devilish old hulk *North Carolina.*"[8]

Nathaniel Hawthorne, when he came aboard in March, was surprised by the comfort afforded by Ericsson's appointments. "It was like finding a palace, with all its conveniences, under the sea . . . [the members of the crew] hermetically seal themselves and go below; and until they see fit to reappear, there would seem to be no power given to man whereby they can be brought to light. A storm of cannon-shot damages them no more than a handful of dried peas."[9] Navy ships had always been domestic spaces where crews ate, read, and sewed during their off hours.[10] The ornamented interior of the "iron home," however, accentuated this convergence of feminine, domestic enclosure with masculine violence.

The *Monitor* thus crossed the borders between surface and depth, comfort and pain, security and risk. A reporter from the *Times* of London reported,

"Her iron deck is not two feet above the water, but on going below she is found to be fitted up with as much luxury below as a yacht."[11] The ironclad recalls the submarine environment of Jules Verne's *Nautilus,* also an elegantly appointed world encased in an aggressive shell. Echoes of the *Monitor,* in fact, appear explicitly in *Twenty Thousand Leagues under the Sea,* published in 1867. Verne's narrator, upon first hearing of the exploits of the *Nautilus,* speculates that it might be a "submarine monitor."[12] As with the *Nautilus,* witnesses to the *Monitor* in battle frequently commented on its menacing aspect, moving freely with no visible human agency.

Comfort and calm continued in accounts of the *Monitor* throughout its operational life, but the experience never regained the novelty and serenity of the first few days at sea, in early March 1862. No moment better illustrates the mood of this short period than Keeler's reaction to the first meal served aboard. The steward, he reported, was drunk and dinner was a "failure": "The fish was brought in before we had finished the soup, and champagne glasses were furnished for us to drink our brandy and vice versa." The steward was dismissed before the vessel left New York.[13] A few days later, Keeler would have wished for these to be his most pressing problems.

"Not Drowned but Stifled"

Between security on one hand and terror and entrapment on the other, lay a range of pleasures and hardships. The *Monitor* crew, despite their pleasing start, never fully adjusted to the new environment, remaining ambivalent throughout their ten-month career on the vessel. The novelty of the sub-merged world thrilled them, but the iron home had significant drawbacks. Darkness, cold, danger, and general discomfort wore on the men. Surviving accounts reveal tensions between faith in new technology and concern for safety, between exciting novelty and threatening uncertainty, between prog-ress and personal experience.

As early as March 18, just two weeks after leaving New York, Keeler la-mented the dehumanizing surroundings: "I'd give a good pair of boots to tread on something besides iron—I am tired of everlasting iron. The clank, clank, clank while I am writing this of the officer of the deck as he paces back and forth on the iron plates above my head, although suggestive of security, is not a good opiate."[14] Though the "clank, clank, clank" refers to a member of the crew, it sounds like the *Monitor's* machinery. Keeler's comment trans-ferred the characteristics of the iron ship, hardness and mechanical motion, onto its human crew.[15] They believed in the progress represented by their new machine, but the *Monitor's* crew also resented the restrictions it put on their own lives.

Those aboard complained about life within the *Monitor's* cramped spaces, the leaky hull, the many problems with machinery, the heat, and the hard, mechanical environment. Frank Butts, for example, who came in October 1862, was thrilled to be assigned to the famous vessel, but he soon found the crew's attitude skeptical: "In the opinion of the crew, . . . a monitor was the worst craft for a man to live aboard that ever floated upon water." Butts quickly came to agree that "the life of a sailor on board this vessel was the most laborious of any in the service . . . I will venture to say that my feet were not dry once in the time I was on board the *Monitor*." Even in port, observed George Geer, the slightest rough water would cause the hatches to leak.[16]

During the summer, conditions aboard the *Monitor* became nearly intolerable, the temperature inside reaching 130° to 150° Fahrenheit. Forced to sleep on deck, Geer complained that "iron is not very soft" but better than inside, for "hell is an icehouse side of this ship."[17] The captain reported to Secretary Welles that the temperature was as high as 156° aboard, adding: "Human endurance has a limit, and it is impossible one should not become exhausted if confined for many hours in such an atmosphere."[18] The crew tried to live outside on deck as much as possible during the summer months, but enemy fire or bad weather could force them to stay below "in our sub-marine cellar," for days at a time. The enlisted men's quarters were less well ventilated than those of the officers, and during the summer men broke down and had to be hospitalized.[19] Captain Jeffers, who took over command of the *Monitor* in the spring, wrote to the Department of the Navy:

> A new and most important defect has developed itself with the warm weather, which demands immediate attention . . . When the weather was cold it was quite warm below, but no inconvenience was felt other than the impurity of the air passing up through the turret; but with the heat of the last ten days, the air stood at 140° in the turret when in action, which, when added to the gases of the gunpowder and smoke, gasses from the fire-room, smoke and heat of the illuminating lamps, and emanations from the large number of persons stationed below, produced a most fetid atmosphere, causing an alarming degree of prostration of the crew.[20]

Keeler described "the bowels of our iron monster . . . densely populated with flies and mosquitoes" and, with his usual flair, recounted, "We lay broiling in our iron box, or cage as it has now become, out of humor with ourselves & the world generally." "This is a species of sea-life," he dryly commented, "of which it is possible for a person to get a surfeit."[21]

Throughout the *Monitor*'s history, living conditions repeatedly arose at the center of debates over the ironclad warship, epitomizing the conflict between engineering and experience that surrounded the new technology of war. Not easily reduced to mechanism, the machine's suitability for supporting life was vague, unscientific, imprecise, unmeasurable, and irreducibly human. John

Ericsson had no purchase on these problems; he only considered those he could calculate and design around. He mocked the crews' complaints about excessive heat, stale air, and the general stifling conditions as womanly weakness, or worse, nostalgia for the comfort of obsolete sailing ships. "As to ventilation, old sailors who have been in these vessels night and day for two years have assured me that no other vessels of war can compare with them."[22] Numerous crew members, however, including Worden, Keeler, Jeffers, Butts, Greene, Newton, and Geer, contradicted this opinion and consistently wrote of the misery belowdecks on the *Monitor*, especially during the summer. Anyone who had been to sea on a monitor knew that discomfort could not be reduced to a matter of personal preference. Living aboard the *Monitor* could make a man sick and unable to fight.

Conditions merely difficult could also become threatening, for the new weapon also presented terrible dangers. The complaints of the crew had a nervous edge; their iron home made them anxious beyond its discomforts. Hawthorne found the enclosure of the *Monitor* not only protecting but also ominous: "In fact, the thing looked altogether too safe; though it may not prove quite an agreeable predicament to be thus boxed up in impenetrable iron, with the possibility, one would imagine, of being sent to the bottom of the sea, and, even there, not drowned but stifled."[23] His comment parallels Keeler's own observation that he was in as much danger from the ship itself as from the enemy. Soon after leaving New York, his insight proved nearly fatally accurate.

Collapse of the Artificial World: The *Monitor*'s First Trip

Early in their first trip, the *Monitor*'s inhabitants confronted the frightening reality of their artificial environment. A major disaster struck that could have ended the career of the experimental craft before it began. Soon after leaving New York on its maiden voyage, the ship hit a gale. The blowers that drew fresh air into the hull protruded above the deck in "stacks" about four feet high. The weather forced water down through the blower stacks and deck hatches. Insufficient caulking of the deck seals exacerbated the problem, and water poured in around the base of the turret, the deck lights, and various other openings in the deck. Executive Officer Samuel Dana Greene wrote to his family, "The water came down under the Tower like a water fall. It would strike the pilot house and go over the Tower in most beautiful curves. The water came through the narrow eye holes in the pilot house with such force as to throw the helmsman completely round from the wheel."[24] Seawater also entered the blowers, leaking onto the leather belts that drove them from the steam machinery. These belts stretched, causing the blowers to fail, in turn causing the engine room to fill with noxious gas from the coal fires.

This event so frightened the crew that a number of detailed accounts survive, as detailed as those of the battle itself. Second Assistant Engineer Albert Campbell drolly recounted:

> We left New York on a beautiful day and . . . the next afternoon we broke both our blower belts which spoiled the draft of our fires and drove all the gas into the engine room. This of course was rather inconvenient, for carbonic acid gas and hydrogen is not calculated to support animal life. I found myself getting weak and lost all consciousness and did not know anymore until I found myself on top of the turret with a couple of engineers lying along side of me, looking more dead than anything else . . . The water poured into the engine room around the smokestacks and ventilators like a miniature Niagara and it was doubtful whether we would float during the night.[25]

To avert suffocation, the engineers opened the engine room doors, but then the entire *Monitor* filled with fumes. Keeler went below to investigate and met one of the engineers, "pale, black, wet, and staggering along gasping for breath."[26]

The engineers were caught in a bind: to make the ship habitable, the machinery had to be repaired, but until the gas was evacuated (i.e., the blowers fixed), no one could work belowdecks. Stimers sent the other engineers up on deck and stayed behind to try to fix the difficulty. He soon found that he "was also getting very much confused in [his] head and very weak in the knees" and had to evacuate as well.[27] The crew of the *Monitor* had stumbled on an unanticipated characteristic of their artificial environment: its ability to support life depended on properly functioning machinery, the blowers. "Then times looked blue I can assure you," Greene wrote.[28] Despite the troubles, the urgent military situation precluded talk of returning to New York.

Eventually, the tugs brought the ship into the lee of the wind near shore. The problem was solved by gathering all the engineers on deck, giving each specific instructions, and sending them below one after the other for short periods to perform small parts of the repair task before the fumes overcame them. Keeler wrote to Anna that he took charge of the engine room while the engineers were recuperating, but none of their accounts mention his assistance (one of the few instances where Keeler's account does not match other documentation).[29] By nightfall the crew reentered the vessel, shaken, sick, and with a new appreciation for the terrible potential of their enclosing home.

Some of these difficulties arose because the urgency of war forced the *Monitor* to rush into battle without a full shakedown cruise. A few weaknesses were easily corrected; Stimers wrote to his wife that he had foreseen the problem with the blowers while the *Monitor* was still in the yard, and "had many discussions with him [Ericsson] about them, but he was very obstinate about it and insisted that four feet was high enough." He thought he could fix the blowers by simply making the ventilation stacks higher to keep out water. Others believed it merely a matter of cutting drain holes in the boxes that enclosed the blower belts so water could not accumulate there.[30] Stimers, courting Ericsson's favor, maintained faith in the vessel, and in a letter to Ericsson after the battle described the trip to Hampton Roads as "a stormy passage which proved this to be the finest sea boat I was ever in."[31] Ericsson replied, "As to the imperfections in the structure, I saw those a long time back; but I also saw how to do them better next time—the insufficient height of the ventilating and smoke trunks—on that point you have floored me."[32] For all his vision and imagination, Ericsson had failed to account for the critical importance of the ventilating shafts. The *Monitor*'s life support systems were a single point failure: when they stopped, the entire system broke down.[33]

Dangerous Novelty

Despite the minor technical nature of these problems, they genuinely shocked the crew—nearly all of them mention the episode when they tell stories of the battle. The dangers of living below the waterline suddenly became frighteningly clear. Several of the men emphasized that the heroism of the *Monitor* crew members lay not in their performance in battle—they were, after all, protected by thick armor—but in their willingness to live in this radically new environment. In Keeler's words, "I think we get more credit for the fight than we deserve—anyone could fight behind an impenetrable armor—many have fought as well behind wooden walls or none at all. The credit, if any is due, is in daring to undertake the trip and go into the fight in an untried experiment and in our unprepared condition."[34] In the hours before the battle, he wrote: "I experienced a peculiar sensation, I do not think it was fear, but it was different from anything I knew before. We were enclosed in what was supposed to be an impenetrable armor—we know that a powerful foe was about to meet us—ours was an untried experiment and the enemy's first fire might make it a coffin for us all."[35] Years later Captain Worden recalled, "Here was an unknown, untried vessel, with all but a small portion of her below the waterline, her crew to live with the ocean beating over their heads—an iron coffin-like ship of which the gloomiest predictions were made, with her crew shut out from sunlight and the air above the sea, depending entirely on artificial means to supply the air they breathe. A failure of the machinery to do this would be almost certain death to her men."[36]

The frequent use in these accounts of the words "experiment," "untried," and "unknown" show that these men were well aware of living on a technological frontier. It threatened not only their safety, but also their very manliness as warriors as they began to question glory's fate in mechanical warfare. As Hawthorne asked, "How can an admiral condescend to go to sea in an iron pot? What space and elbow-room can be found for the quarterdeck dignity in the cramped lookout of the Monitor, or even in the twenty-feet diameter of her cheese-box?"[37] It takes courage to follow an untried path, especially into the hostile environment of the ocean, and it takes courage to stake one's life on new machinery, especially in the face of battle. The machine's mediation of the battle with nature at sea paralleled its mediation of the battle between enemies at war—one reason, perhaps, that it became such a compelling icon. But at what point does the mediating machine, rather than its human operators, become the agent of war and hence the hero?

The rest of the *Monitor*'s journey from New York, although brief, was not uneventful. Water continually poured in through the base of the turret, threatening to disable the blowers or put the fires out. The ropes jumped off the steering wheel and had to be replaced. Lieutenant Greene, on watch through the night, thought dawn might never come: "I certainly thought old Sol had stopped in China and never intended to pay us another visit."[38] Much of the next day was calm, although water still flowed in, doing nothing to relieve the exhausted and uncomfortable crew. The hawser that connected the vessel to the towboat parted and had to be repaired. In the afternoon the *Monitor* passed Cape Henry and the crew heard firing from Hampton Roads. A pilot came aboard and brought them the bad news: they had missed the *Virginia* by one day. That evening, as the *Monitor* entered the roads, they could see a ship on fire. The Confederate ironclad had already arrived, wreaking havoc on the Union blockade.

Chapter Five

The Battle of
Hampton Roads

 Fine clear weather. At 7:20 got underweigh, all hands to quarters. 8:20 opened fire on the Merrimack, from that time until 12, constantly engaged with Merrimack. 12:30, rifled shell struck the pilot house, severely injuring commander Worden. Merrimack hauled off in a disabled condition. 2pm Captain Worden left for Fort Monroe.

Ship's log, USS *Monitor,* March 9, 1862

As the *Monitor* appeared on the stage of its greatest performance, the crew came to realize that conditions aboard affected more than their letters and casual observations. They affected the experience of combat and the outcome of the major battle that lay just ahead. During this conflict, sealed isolation in the hull indeed withstood the force of the ocean and the enemy's exploding shells, but afterward it succumbed to the world's prying gaze. The *Monitor* crew became heroes of the Union, publicly visible because of their very invisibility in the machine.

The crew of the *Monitor* not only experienced the effects of these image politics; they actively participated in shaping them, welcoming visitors and reporters into their iron home. One of those visitors was Nathaniel Hawthorne, who saw in the *Monitor* implications not only for the immediate situation of the war, but also for the meaning of heroism in an age of technology. His insights proved accurate, for the navy itself realized that the new weapon's visibility, while exaggerating its strength, could also expose its weaknesses. The *Monitor,* once announced to the public, became a secret weapon, its heroes hidden from public view.

Appearance of the *Virginia*

The water hisses & boils with indignation as like some huge slimy reptile she slowly emerges from her loathsome lair.

William F. Keeler to Anna Keeler,
on seeing the *Virginia*, May 7, 1862

During the *Monitor*'s development, rumors about the conversion of the *Merrimack* drifted north through intelligence channels. At the Gosport Navy Yard at Norfolk, Virginia, the *Merrimack*, rechristened the *Virginia*, was being raised and rebuilt as an ironclad (although many Northerners continued to refer to it as the *Merrimack*). Even with the industrial facilities of Gosport at their disposal, however, Southern builders were hampered by poor equipment and a shortage of skilled labor. They could only recondition the original engines of the *Merrimack*, which had been condemned for replacement even before the ship sank. The Tredegar ironworks in Richmond rolled two-inch-thick armor plate for the *Virginia*'s sides. The new ship, 262 feet long, had an armored casemate with angled sides 170 feet long protruding above the water. The armor's weight gave the vessel a deep, 22-foot draft, severely limiting its freedom of movement (and evidence suggests that that vessel was just as uncomfortable as the *Monitor* to live aboard, although its crew inhabited it for shorter periods).[1] Despite its makeshift nature, the project instilled fear in Washington. By the time the *Monitor* was completed and under way, the fear had turned to panic. All expected the *Virginia* to challenge the Yankee blockade. If successful, it was feared, the vessel could steam up the Potomac and shell Washington itself.

Propitiously, the *Virginia* arrived at Hampton Roads on March 8, 1862, just one day before the *Monitor*. The result was disaster for the Union fleet, which was caught unprepared and doomed by its own mistakes. The *Virginia* fought the Union fleet with virtual impunity, ramming and sinking the corvette *Cumberland* and killing 121, about a third of its crew. Thomas Oliver Selfridge, Jr., the officer of the deck (who soon after briefly commanded the *Monitor*), recalled "a scene of carnage unparalleled in the war."[2] The *Virginia* also set afire the *Congress*, which surrendered and later exploded, with 240 dead, more than half of its complement. Included among the dead was Lieutenant Joseph Smith, son of Commodore Smith, who had overseen Ericsson's building of the *Monitor* for the navy. By nightfall the Confederate ironclad withdrew, largely intact (though minus its ram), leaving a third Union frigate, the *Minnesota*, aground in an ebbing tide, stuck at least until morning, when it would be an easy target.

The circumstances of the day's events made the fighting a less-than-objective comparison of wooden versus iron ship technologies. Nonetheless, the public and the navy saw the events of March 8 as a vivid demonstration of the superiority of ironclad to wooden vessels and gave new credibility to the ancient tactic of ramming. The bad news quickly reached Washington, and the president summoned his advisors, including John Dahlgren and Navy Secretary Welles, who recalled the mood as "gloomy . . . the most so I think of any [time] during the rebellion." Lincoln had a special interest in new weapons, which he pursued with a hobbyist's enthusiasm during the war, so he appreciated the implications of the *Virginia*'s victory as well as the uncertain state of the *Monitor*.[3] Welles assured Lincoln that the new *Monitor* was on its way and that the capital was safe because the deep draft of the Confederate ironclad would prevent it from passing up the river. Even so, according to Welles, all through the evening Lincoln "went repeatedly to the window and looked down the Potomac—the view being uninterrupted for miles—to see if the *Merrimack* was not coming to Washington."[4]

That same evening, March 8, the *Monitor* arrived at Hampton Roads, her crew exhausted by the difficulties of their first trip. It was soon after they arrived that the *Congress* blew up. "A grander sight was never seen," wrote Lieutenant Greene, "but it went straight to the marrow of our bones."[5] The stage for the next day's battle was set, the strategy clear: the *Monitor* would defend the stranded *Minnesota* from the *Virginia*.

Unique circumstances leading up to the battle between the two ironclads added to the drama and spectacle of the encounter. Through different routes, two separate navies had developed ironclad technology. The South had raised a burned Union warship, stripped the rigging from its hull, and built an angled, ironclad battery down its length, supporting ten guns total. The North, in contrast, had built a radically new design from scratch with an iron hull and two guns. The two warships arrived on the scene within a day of each other. And that one day made a deadly difference. Any number of problems could have further delayed the *Monitor,* bringing even greater disaster for the Union navy. Although remarkable, this timing was no coincidence. The *Monitor* had been hurried in direct response to the anticipated conversion of the *Merrimack*.

The Naval Amphitheater

The ironclads' encounter in Hampton Roads was intimately connected to the larger evolution of the war. Hampton Roads was a key strategic point, both for enforcing the Union blockade and for supporting land operations. Here the James, Elizabeth, and Nansemond Rivers meet Cheseapeake bay at the tip of the peninsula formed by the James and York Rivers. General McClellan had chosen this peninsula as his avenue to attack Richmond, in order to have logistics and fire support from the rivers that run up both its sides. McClellan intended to land his army of nearly one hundred thousand at the massive Fortress Monroe, "the Gibraltar of Chesapeake bay," at the entrance to the Roads and then to transfer tens of thousands of troops through the Roads to the peninsula to begin the campaign. Had the Peninsula Campaign succeeded, it would have significantly altered the outcome of the war. Instead, it received a severe blow from the *Virginia*'s appearance. "The performance of the *Merrimack*," wrote a cautious McClellan, "place[s] a new aspect upon everything and may very probably change my old plan of campaign just on the eve of execution."[6] Effectively closing off access to the James River divided McClellan's naval support in two. Moreover, instead of heading up the York River, the navy stayed in Hampton Roads to defend against the *Virginia*. Thus the Confederate ironclad virtually eliminated naval involvement in the Peninsula Campaign for several critical weeks.[7] No technological sideshow, the encounter between the *Monitor* and the *Virginia* was strategically central to the Peninsula Campaign and to the war.

The geography of Hampton Roads also added to the theatrics of the battle. A natural public event, the encounter between the *Monitor* and the *Virginia* took place in full view of tens of thousands of troops from both sides, as well as a number of British and French warships in Hampton Roads as observers. Keeler captured the sense of spectacle when he described the emergence of the *Virginia* as "like some huge gladiator just entering the vast watery arena of the amphitheater." Similarly, historian William Still, Jr., has called Hampton Roads a "natural naval amphitheater" because of the numerous coastlines that extend from the water in all directions.[8] Had this battle occurred in bad weather, at sea, or in an isolated location, it would not have seized the American imagination in the way it did. The naval "revolution" was a public event.

On the morning of March 9, Keeler and his friend the surgeon tarried on deck for a close look at the *Virginia* as it approached. They saw a puff of smoke, and shortly thereafter a shell sailed over their heads and slammed into the side of the *Minnesota*. The battle was beginning; Keeler and his friend scampered below. "We did not wait [for] a second invitation but ascended the tower & down the hatchway . . . The iron hatch was closed over the opening & all access to us cut off."[9] Then the *Monitor* and the *Virginia* engaged for more than four hours at close range, sometimes even colliding and brushing against each other as the *Virginia* attempted to ram the smaller vessel. The *Monitor* fired solid shot, the *Virginia* exploding shells. Observers and participants were amazed to see both shot and shells literally bouncing off the iron combatants, "with no more effect, apparently, than so many pebble stones thrown by a child."[10] The *Monitor* took twenty-one or twenty-two hits, mostly in the turret, but suffered only minor damage. The *Virginia* had most of its extremities shot away. The crew's inexperience rendered the *Monitor*'s

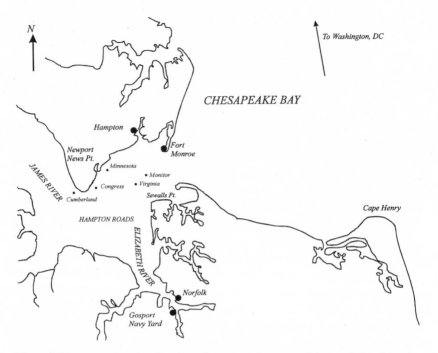

Map of Hampton Roads, showing the approximate positions of the ships involved in the battle.

firing haphazard and tactically incoherent. Reloading alone took seven to eight minutes. Furthermore, the Dahlgren guns had been de-rated with less than their maximum charge of powder, because of the fear of bursting their breeches. Firing with fifteen pounds of powder guns that would later support fifty, the *Monitor*'s gunners frequently failed to penetrate the enemy's armor. In midafternoon the two vessels disengaged with no fatalities and only one injury.

That one injury, however, ended the encounter: a shell explosion blinded the *Monitor*'s captain by blasting into his eyes through the slit in the armor of the pilothouse (wheelsman Peter Williams later won the Medal of Honor for keeping his hand on the wheel during the change of command). What happened next remained the single most controversial question about the battle for the *Monitor* crew. After some delay, which reports put at anywhere from ten to thirty minutes, Lieutenant Greene took over command of the ship. The vessel drifted off, he maintained, in the confusion resulting from the change of command.[11] Other crew members (mostly enlisted men) later accused Greene of cowardice because he did not pursue the *Virginia* and destroy it. The officers generally defended his conduct.[12] A Southern paper reported on March 10 that the *Virginia* "plunged full tilt at the *Ericsson* [the *Monitor*], causing the Yankee iron monster to head instantly for Old Point, with all hands at pumps, in a supposed sinking condition." Some observers said the *Virginia* withdrew because of damage, while the *Monitor* remained behind to defend the *Minnesota* as ordered. Either way, the *Virginia* headed back up the river to Norfolk and the *Monitor* did not pursue. Both sides claimed victory.[13] For the Union, the *Monitor* had saved the blockade fleet from the threat of the *Virginia* and achieved its tactical goal of protecting the *Minnesota*. The Confederacy saw the *Virginia*'s destruction of the *Congress* and the *Cumberland* as evidence of its triumph, in addition to what they interpreted as the *Monitor*'s abandoning the field of battle in Hampton Roads. Because it closed the James River to Union traffic, strategic arguments favor the *Virginia,* whereas tactical arguments favor the *Monitor*.[14] Nonetheless, the battle ended with stalemate, ambiguity, and controversy: perfect ingredients for legend.

Command and Confusion aboard the *Monitor*

Well-informed naval officers are aware that Worden failed to sink the Merrimack *at Hampton Roads because he could not personally control the firing and at the same time direct the steering of his vessel from a point enabling him to observe properly the movement of his antagonist . . . excepting the omission to place the pilot-house on the top of the turret, the original* Monitor *was a perfect fighting machine.*

John Ericsson, 1876

Who "won" the battle, who should have won the battle, and who would have won a subsequent battle are three unanswerable questions, sources of endless debate by historians and observers. They concern us here less than the human activity aboard the *Monitor* during battle and how conditions aboard affected its crew and their performance. The crew's experience aboard the *Monitor* differed significantly from what Ericsson had foreseen when he designed the vessel. Machinery imposed constraints on the crew, altered their relationship to each other, and affected the battle's outcome.

The unique structure of the *Monitor* divided the crew and interrupted the chain of command. Captain and pilot occupied the forward pilothouse while, in the turret, the executive officer directed fire with the gun crew. None doubted that the gun crew was well protected inside the turret; Chief Engineer Stimers wrote to his father of the remarkable effect of the *Virginia*'s shot: "My duty was to turn the turret to bring the guns to bear on the enemy, the lieutenant of the ship sighting them . . . The crash against the turret was tremendous when their heavy shot struck it and if a man happened to lean against the inside of where a shot struck it knocked him down and stunned him for a couple of hours."[15] The surgeon confirmed the effect, reporting two concussion injuries.[16]

Such protection, however, came at the cost of the gun crew's disconnection from the enemy, and from the commander. A speaking tube connected the pilothouse and the turret, but it was out of commission during the battle (nearly all crew members' accounts mention the broken speaking tube, although none explains how such a simple device failed). Thus, the captain and the pilot could not communicate directly with the gunnery crew during the fight with the *Virginia*. The lieutenant in charge of the gun crew, Executive Officer Samuel Dana Greene, reported the difficulty of performing his duty from the armored position:

> The effect upon one shut up in a revolving drum is perplexing, and it is not a simple matter to keep the bearings. White marks had been placed upon the stationary deck immediately below the turret to indicate the direction of the starboard and port sides, and the bow and stern; but these marks were obliterated early in the action. I would continually ask the captain, "How does the *Merrimack* bear?" He replied "On the starboard-beam," or "On the port-quarter," as the case might be. Then the difficulty was to determine the direction of the starboard-beam, or port-quarter, or any other bearing.[17]

Greene later recalled, "The drawbacks to the position of the pilothouse were soon realized . . . Keeler and Toffey [Keeler's clerk] passed the captain's orders and messages to me, and my inquiries and answers to him, the speaking tube from the pilothouse to the turret having been broken early in the action. They performed their work with zeal and alacrity, but, both being landsmen, our technical communication sometimes miscarried." (Keeler later objected to Greene's characterization of him as a "landsman," citing his two years at sea prior to the war.)[18] The new command configuration created by the *Monitor's* physical structure interrupted "technical communication" between the captain and his crew, leading to a crisis of control in combat. For years after, the debate centered on Greene's performance, but he placed the blame on the machinery: "The situation was novel: a vessel of war was engaged in desperate combat with a powerful foe; the captain, commanding and guiding, was enclosed in one place, and the executive officer, working and fighting the guns, was shut up in another, and communication between them was difficult and uncertain."[19] If we think of the *Monitor* as a living organism going into battle, it had intermittent and unreliable connections between its eyes, its brain, and its fists.

Paymaster Keeler made those connections. The skilled communicator spent the battle below deck conveying information between Greene and Captain Worden. He probably qualified for this job because of his semi-outsider status; everyone else aboard had specific duties during combat, but the paymaster had no accounts to keep. Keeler, even more removed from the *Virginia* than those in the turret, reported the frustrating confusion of experiencing the battle from below the waterline:

> With the exception of those in the pilot house and one or two in the turret, no one of us could see her [the *Virginia*]. The suspense was awful as we waited in the dim light expecting every moment to hear the crash of our enemy's shot . . . Mr. Greene says "Paymaster, ask the captain if I shall fire." The reply was "Tell Mr. Greene not to fire till I give the word." . . . Below we had no idea of the position of our unseen antagonist, her mode of attack, or her distance from us, except what was made known through the orders of the captain. "Tell Mr. Greene that I am going to bring him on our starboard beam close alongside."[20]

Keeler later received a commendation from Secretary Welles for "carrying the orders between the pilot house and the turret. Giving them always in a cool distinct manner that added greatly to the complete understanding between these two positions."[21]

In this platitude Welles gave the impression of greater control and understanding than the crew had reported. Indeed, themes of control and maneuverability appear frequently in observations of the battle. The *Baltimore American*, for example, described the battle as witnessed by an observer ashore, conveying the sense that the crew had total command of the vessel: "The *Monitor* ran round the *Merrimack* repeatedly, probing her sides, seeking for weak points, and reserving her fire with coolness, until she had the right spot and the exact range, and made her experiments accordingly . . . the *Merrimack* turned toward Sewell's point and made off at full speed."[22] Because the *Monitor* had no visible crew, no apparent means of propulsion, and barely any unprotected moving parts, its movement—freer and quicker than a traditional warship, even a steam-powered one—seemed miraculous. These images contributed to the sense of the *Monitor* as a willful machine, something slightly supernatural, and to the perceived superhuman status of its crew.

After the battle, the famished and exhausted crew sat down for lunch. Assistant Secretary of the Navy Gustavus Fox, whom Ericsson and his allies had worked so hard to convince in the preceding months, had witnessed the battle from the *Minnesota*. The spectacle completed Fox's conversion, which he quickly telegraphed to Secretary Welles in Washington. Then Fox came aboard and gave his commendation, "Well, gentlemen, you don't look as though you were just through one of the greatest naval conflicts on record." Captain Worden was removed for treatment of his injuries, and Lieutenant Greene became temporary captain. "I had been up so long, had had so little rest, and had been under such a state of excitement that my nervous system was completely run down . . . My nerves and muscles twitched as though electric shocks were continually passing through them and my head ached as if it would burst . . . I laid down and tried to sleep, but I might was well have tried to fly."[23] Fox sent a man aboard, William Flye, to function as First Lieutenant and help the inexperienced Greene with his new responsibilities.[24] Keeler, feeling pensive, walked out on the deck, littered with fragments of exploded shells. The debris reminded him of the value of the *Monitor's* armor. He collected a few fragments as souvenirs and sent one home to his wife and another to President Lincoln, "with the respects of the officers of the *Monitor.*"[25] He would soon pay those respects in person.

Public Response to the *Monitor's* "Victory"

News of the battle spread swiftly. As the *Monitor* docked, its crew were treated as heroes by the navy, the army, and the public. Keeler wrote of the *Monitor's* reception, "We passed along close to Newport News. The whole army came out to see us, thousands and thousands lined the shore, covered the vessels at the docks & filled the rigging. Their cheers resembled one continuous roar." Union leadership hailed the *Monitor's* performance as a victory, spelling the end of traditional navies. On March 13 Senator James Grimes, who had guided the ironclad bill through Congress, announced to Congress that wooden vessels were now "harmless and helpless," that ironclads could steam into any harbor defended by even the "strongest stone fortifications," and that "we can now commence the creation of a proper navy on a footing of comparative equality with all the naval powers of the world."[26] The mechanical image of the small, geometric *Monitor* had tremendous appeal, way beyond that of ironclad gunboats on the western rivers. Starved for good news early in a slow and bloody war, the Northern public welcomed the *Monitor* as the perfect technological solution to the Union's problems. A "monitor craze" swept the North: consumers were offered *Monitor* cigars, suits, hats, playing cards, songs, even *Monitor* flour. *Monitor* builder Cornelius Bushnell was quickly elected to the Connecticut State Legislature.[27]

Tied up at Hampton Roads, the *Monitor* attracted important visitors from Washington and elsewhere, eager to see firsthand the famous war machine. "Everything you see here is epaulette, shoulder strap, buttons & Swords," Keeler observed. Visitors included the Prince de Joinville of France, a leading figure in the earlier debates about naval armament, who had witnessed the battle. General McClellan came, the nineteenth of his rank to visit, as did "any quantity of foreign nobles, counts, &c," including a number of Swedes, interested in the product of their countryman John Ericsson. "I do not believe there is a place in the United States," wrote Keeler, "where we could see as many of the noted men of the country—Naval, Military, and Civil—as just where I am. Even in Washington you would have to hunt them up, but here they all center at this point."[28] In part because of his skill with language, and because he frequently had few duties, Keeler became the tour guide for the *Monitor* in these "fresh irruptions of visitors." Selected to tour visitors around the ship in "nine cases out of ten," Keeler quickly tired of all the ushering and exhibiting, even proposing to write a guidebook to hand to the visitors as they arrived so they might show themselves around. He narrated the book's contents to Anna: "The turret, gentlemen . . . Here is where we were hit by a 100 lb. Percussion shell from the *Merrimack*."[29]

On May 7, two months after the battle, President Lincoln came aboard the *Monitor* with Secretary of Treasury Salmon P. Chase and Secretary of War Ewin M. Stanton. Lincoln had come into the field to personally direct the Peninsula Campaign for a short time. Keeler noted, as did others, that the president seemed detached and burdened, "in strong contrast with the gay cortege by which he was surrounded." "As the boat which brought the party came alongside every eye sought the *Monitor* but his own. He stood with his face averted as if to hide some disagreeable sight. When he turned to us I could see his lip quiver & his frame tremble with strong emotion & imagined that the terrible drama in these waters of the ninth [eighth] & tenth [ninth] of March was passing in review before him." When meeting Keeler, Lincoln was pleased to find someone from his home state, and the paymaster was impressed that he turned down the offer of a drink. The president, with an educated interest in military technology, then toured the vessel.

> He examined everything about the vessel with care, manifesting great interest, his remarks evidently shewing that he had carefully studied what he thought to be our weak points & that he was well acquainted with all the mechanical details of our construction . . . Most of our visitors come on board filled with enthusiasm & patriotism ready, like a bottle of soda water, to effervesce the instant the cork is withdrawn, but with Mr. Lincoln it was different. His few remarks as he accompanied us around the vessel were sound, simple, & practical, the points of admiration & exclamation he left to his suite.[30]

A boat full of congressmen followed the president aboard; but as they unloaded, the signal came to prepare for combat. The *Virginia* had been spotted coming around Sewell's point. Although a false alarm, the news cleared the vessel of visitors.

For all the bother of visitors, Keeler always enjoyed showing the ship to women. While admirals and congressmen met in the wardroom, the paymaster toured their wives and daughters around the ship. He notably included them, although rarely by name, in his description of the groups: "Vice President Hamlin & lady, Senator Hale & lady &, I believe, two or three daughters."[31] As he had with his early encounters with women in the Brooklyn Navy

Yard, Keeler gave these experiences special attention in his letters to Anna. In his own recounting, Keeler's female companions responded to his tours with a flirtatious charge:

> A Mr. Wall connected with the Navy Department in Washington introduced me to his wife with the request that I would take charge of her for a time—as she was young, handsome, & intelligent of course I couldn't refuse. I asked here if she had been in the turret to see the guns. "Oh yes," she said, "& kissed them too. I feel as if I could kiss the deck we stand on," &, continued, one of her female friends who was standing near, "I would like to kiss all who were on board during the fight if I thought they would let me." I don't know but that I should have taken advantage of this fit of enthusiasm if I hadn't thought it might by some accident have reached your ears.[32]

At sea, the artificial world of the *Monitor* was exclusively male. But that realm defined the *Monitor* no more than did John Ericsson's drawings. Keeler's sensitivity to the presence of women in his military world highlights their role in making the vessel into a legend and the crew into heroes. Social status followed glory in war. For Keeler, its hallmark was this new and exciting interaction with women. Without their attention, their gazes, their curiosity, Keeler would not have enjoyed the same luminosity of fame. These attentions formed the public, social counterpart to Keeler's private world, where letters represented life aboard the *Monitor* specifically for Anna, and for her to convey to others. For Keeler, the woman in his life validated the experience of war. In her absence, other women authenticated his heroism. To the extent that we view the history of the *Monitor* through Keeler's eyes, we see it through the eyes of Anna and of Keeler's female visitors. Women, as observers and readers, the prototype eyes of history, participated in both the daily experience and the public display of the machine.

Hawthorne's Detached Inspection

Keeler usually informed Anna of his famous visitors, although one escaped mention. About March 30, three weeks after the battle, Nathaniel Hawthorne visited the *Monitor* in Hampton Roads. Hawthorne wrote an article about his trip, "Chiefly about War Matters by a Peacable Man," which appeared in the July 1862 issue of *Atlantic Monthly;* it included a lengthy section on his visit to the *Monitor* and his reactions to the strange new machine. The article deeply affected its audience, as writer and editor George William Curtis wrote: "What an extraordinary paper by Hawthorne in the 'Atlantic,'! It is pure intellect, without emotion, without sympathy, without principle. I was fascinated, laughed and wondered. It is as unhuman and passionless as a disembodied intelligence."[33] Before examining Hawthorne's writing on the *Mon-*

itor, however, let us briefly consider his prior interest in the industrialization of American society and the circumstances that brought him to Hampton Roads.

"Chiefly about War Matters" was Hawthorne's first writing about war, but the *Monitor* did raise questions he had dealt with before in other arenas. The storyteller from Salem, Massachusetts, had become famous for his novels *The Scarlet Letter, The House of Seven Gables,* and *The Blithedale Romance* and for his short stories, published in the decade or two before the Civil War. Several of Hawthorne's stories focus on scientists. "Dr. Heidegger's Experiment" features a doctor who experiments on his friends with a formula for the fountain of youth. "Rapaccini's Daughter" tells of a botanist who breeds his daughter from poisonous flowers. These characters were not the mechanics, inventors, or tinkerers of early-nineteenth-century America. Rather, they were romantic, alchemical scientists obsessed with knowledge in a Faustian mode, and sometimes hucksters or pseudoscientists bent on exploiting falsehood for profit. (Scientists were not linked to industrialization until later in the century.) Another story, "The Artist of the Beautiful," depicts a young watchmaker who builds a mechanical butterfly for the woman he loves. Even here, Hawthorne's theme is not machinery per se, but love, perfection, and loss, his artificer less the mechanic than the artist and romantic idealist.

Hawthorne had also written about the dangers of industrialization, and these ideas informed his response to the *Monitor.* The story "The Celestial Railroad" retells John Bunyan's allegory of Christian's pilgrimage. Instead of journeying to heaven on foot, though, here pilgrims ride to the Celestial City on a railroad that eases their burdens. In this world, machines effortlessly convey morality and wisdom, but railroad iron is forged in the infernal regions, inhabited by "unlovely personages, dark, smoke-begrimed, generally deformed, with misshapen feet, and a glow of dusky redness in their eyes." In "The Celestial Railroad" (and in other stories, such as "Ethan Brand"), Hawthorne represented changes wrought by industry and manufacturing with devices of allegory and irony.[34]

Nevertheless, Hawthorne's main interest was not mechanics but "mechanism," the increasingly instrumental, pragmatic, and commercial orientation of the country. American culture, flush with material improvements and conveniences, seemed to Hawthorne blindly barreling toward a dark future. The rise of railroads and factories threatened the Jeffersonian, pastoral ideal of a pure America free from Europe's degrading influences. What Leo Marx wrote of Thoreau also applies to Hawthorne (and, for that matter, to Melville as well); he used "technological imagery to represent more than industrialization in the narrow, economic sense. It accompanies a mode of perception, an emergent system of meaning and value—a culture."[35] Social and technological change was not a new subject to Hawthorne. To him, the train whistle he heard in the Concord woods signaled the vanguard of an industrial future, bringing "the noisy world into our slumbrous peace."[36] The ominous, black warship seemed to embody the dark side of those same forces, forces that led the country into a new age and then into cataclysm.

By the time he visited the *Monitor,* Hawthorne fit under Keeler's heading of "noted men of the country" and had long considered the changes wrought by new mechanism and industry. But Hawthorne always had a peculiar relationship to current events and mainstream culture. He spent much of his early life in solitary writing, although he had some flirtations with politics (President Franklin Pierce was a close friend). He believed the war justified but thought the North should defeat the South and then let it go. "Amputation seems to me much the better plan," he wrote in 1861.[37] Nevertheless, the outbreak of war thrilled him, as it did many of his generation; he repeatedly expressed regret at being too old to serve, an excitement that made its way into his fiction.[38] Yet, on a deeper plane, Hawthorne was out of touch with the mood of the country in its darkest hour, his inherent isolation exaggerated by having spent the seven years prior to the war in England as a consul. Although he shared the country's early enthusiasm, as it got under way the war alienated and depressed him. Hawthorne's "Chiefly about War Matters" benefits from this detachment, for it freed his observations of the *Monitor* from the public celebration that followed Hampton Roads.

In early 1862 Hawthorne's health was poor, and his friend Horatio Bridge invited him to Washington, thinking a trip south would improve him. Hawthorne declined, until James T. Fields, his publisher, suggested that he write about the experience for Fields's *Atlantic Monthly,* then the nation's leading literary journal. Hawthorne agreed, and he soon left New England by train, headed south to war, in the company of Fields's partner, William Ticknor. It was March 10, 1862, the day after the Battle of Hampton Roads. The trains crawled with curious soldiers in search of newspapers from which to read accounts of the battle. Hampton Roads was the furthest south in his country that Hawthorne would ever venture.

"There is no remoteness of life and thought, no hermetically sealed seclusion, except, possibly, that of the grave, into which the disturbing influences of the war do not penetrate," begins "Chiefly about War Matters," which recounts Hawthorne's trip. To this morose romancer, war was revealing itself as a force exceeding allegory. "There is a kind of treason in insulating one's self from the universal fear and sorrow, and thinking one's idle thoughts in the dread time of civil war." As Hawthorne and Ticknor moved south, the air became warmer and the atmosphere more martial, until by the time they reached Washington little was apparent save soldiers and guns. Hawthorne recorded the minute sensations of descending into a war zone. The land bore prominent signs as armies left destruction in their wake. "Around all encampments, and everywhere along the road, we saw the bare sights of what had evidently been tracks of hard-wood forest, indicated by the unsightly stumps of well-grown trees, not smoothly felled by regular axe-men, but hacked, haggled, and unevenly amputated, as by a sword or other miserable tool, in an unskillful hand. Fifty years will not repair this desolation."[39] His descriptions are irreverent, apolitical, and frequently at odds with common sense. He mocks the vaingloriousness the country would have to suffer: "Civil life will be colored by veterans' boasts for a generation." "One bullet-headed general will succeed another in the Presidential chair."[40] Despite points of enthusiasm, writes Randall Stewart, "the prevailing mood of the essay is one of perplexity with respect to war issues—a perplexity that was increased by a disposition to see both sides of the conflict."[41] This perplexity, however, as well as an ironic distancing, enabled Hawthorne to see the ironclads as portents as well as artifacts, harbingers not merely of a new age of naval technology but also of novel mechanisms of strife and new ways for machines to congeal human conflicts.

After arriving in Washington, Hawthorne ventured to Hampton Roads, part of a delegation sent by the secretary of war to inspect the forces and report back. There they examined Fortress Monroe and its defenses and went aboard several Union vessels in the area, including the steam frigate *Minnesota* and the *Monitor*. Aboard the former, which, like the original *Merrimack,* contained a steam engine but still relied primarily on sail power, the delegation was "shown every part of her, and down into her depths, inspecting her gallant crew, her powerful armament, and her furnaces."[42] Despite the impressive scenes, Hawthorne saw the vessel as "as much a thing of the past as any of the ships of Queen Elizabeth's time, which grappled with the galleons of the Spanish Armada." More obsolete than its equipment were its people and their affectations. Social change had to accompany mechanical invention.

> On her quarter-deck, an elderly flag-officer was pacing to and fro, with a self-conscious dignity to which a touch of the gout or rheumatism, perhaps contributed a little additional stiffness. He seemed to be a gallant gentleman, but of the old, slow, and pompous school of naval worthies, who have grown up amid rules, forms, and etiquette which were adopted full-blown from the British navy into ours, and are somewhat too cumbrous for the quick sprit of to-day. This order of nautical heroes will probably go down, along with the ships in which they fought valorously and strutted most intolerably.[43]

Aboard the *Monitor,* it is likely that Keeler showed the writer around his ship. Unfortunately, neither man mentions the other, but we do know that Keeler spent these days busily entertaining the vessel's many guests. Hawthorne had embarked on his tour of the war zone at the request of Horatio

Bridge, his college friend who served as paymaster general of the navy. Bridge was the highest-ranking man in the job Keeler himself performed and accompanied Hawthorne to Hampton Roads. As paymaster, Keeler had regular correspondence with Bridge, his superior. Possibly, then, Hawthorne had special interest in the articulate paymaster. We shall never know for sure if the two men spoke, but their relationships to Bridge and the similarity of their views of the *Monitor* strongly suggests that they did.

Hawthorne's description of the *Monitor,* "the strangest-looking craft I ever saw," echoes those of Keeler and the crew: "It could not be called a vessel at all; it was a machine." The author articulated the uncanniness of the vessel that lived below the waves and had no external evidence of human activity: "It was ugly, questionable, suspicious, evidently mischievous—nay, I will allow myself to call it devilish." His image of the vessel under way presents the external counterpart to Keeler in his cozy wardroom: "The billows dash over what seems her deck, and storms bury even her turret in green water, as she burrows and snorts along, oftener under the surface than above." Actually, Hawthorne probably did not see the ship under way, but he could not resist his imagination. "The singularity of the object has betrayed me into a more ambitious vein of description than I often indulge; and, after all, I might as well have contented myself with simply saying that she looked very queer."[44]

Hawthorne saw the *Monitor* as foretelling a new mode of war. He recognized that the transformation would bring expertise and experience into conflict, and he found that conflict central to the *Monitor*'s significance. In his sardonic but thoughtful piece, Hawthorne articulated a vision of mechanical warfare, imagining not only a reduced human role at the face of battle (as with the *Monitor*), but also a time when human beings would become absent from the fight altogether:

> There will be other battles, but no more such test of seamanship and manhood as the battles of the past; and, moreover, the Millennium is certainly approaching, because human strife is to be transferred from the heart and personality of man into cunning contrivances of machinery, which by-and-by will fight our wars with only the clank and smash of iron, strewing the field with broken engines, but damaging nobody's little finger except by accident. Such is obviously the tendency of modern improvement.[45]

With its transference of strife from humans to machines, Hawthorne's vision would today be called cybernetic. Nonetheless, he saw that the conversion was a social change as much as a technical one, predicting the decline of the traditional "order of nautical heroes." Recall his question, "How can an admiral condescend to go sea in an iron pot?" Again, we are reminded of Rosalind Williams's observation that literary figures of the underground represent visions of a technological future. So, too, for the enclosed, mechanical spaces of the *Monitor:* Would futuristic machinery eliminate human involvement from war? Or would the *Monitor* usher in a new breed of heroes?

"War in Reality"

Ironically, whereas Hawthorne envisioned a war completely free of warriors and heroism, in the weeks after the battle Keeler and his companions were experiencing the full flush of military glory. The visitors to the ship itself all had high social status, but not only the elite hailed the crew. Men of all ranks had witnessed the battle from shore. They knew the damage the unchecked *Virginia* had caused and the value of the *Monitor's* protection. When the navy chartered a small steamboat to tend the vessel and run errands around the bay, Keeler went ashore to get provisions. "You cannot conceive of the feeling there," he remarked; "the expressions of gratitude & joy embarrassed us they are so numerous."[46] Albert Campbell was thrilled to find that "people ashore could not say enough in our praise and would not take our money for anything."[47]

Keeler's excursions ashore highlighted the social stratification of war experience. In this, his first observation of war on the ground, he saw that the docks and streets were filled with troops, wagons, artillery, and equipment as the Peninsula Campaign amassed forces for its movement toward Richmond. The rush and novelty excited him, but Keeler also realized how atypical had been his own experience. Seeing a long line of combat-worn infantry unloading from a transport, he noted that "they were mostly young looking men, hardy & bronzed from constant exposure . . . I could think of nothing but a lot of overloaded pack horses as they passed along with their shoulders stooping under their loads . . . they hadn't that lively air & mien & the military gait you see in soldiers on parade." Suddenly the pomp and power to which he was becoming accustomed juxtaposed with a grittier image: "This is war in reality, divested of its paint and feathers, and I must say viewed in this light it lacks much of its poetry and romance." George Geer had a similar reaction when observing a steamboat loaded with troops. In future letters Keeler would speak only sardonically about the "pomp & circumstance of glorious war" as he was increasingly exposed to its human faces, military and civilian.[48]

What affected Keeler about these scenes was their ironic contrast with his own experience. He fought behind thick armor plates and he never saw the enemy. Recall Keeler's initial thought, when the *Monitor* was under construction, that "there is not danger enough to give us glory" and Hawthorne's note that "even heroism—so deadly a gripe is Science laying on our noble possibilities—will become a quality of very minor importance." Yet the *Monitor* crew, by virtue of the public image and the public fight of their vessel, had had the opposite experience as warriors. They stumbled through one short, indecisive battle and suddenly found themselves the heroes of the entire Union. Keeler would soon realize, however, that the giddy certainty of the days after the fight could not last. Now, with the *Monitor*'s reputation for invulnerability and its public image established, the *appearance* of superiority rather than its exercise would become the vessel's primary source of value.

Chapter Six

Iron Ship in a
Glass Case

April–September 1862

I wish the public could see all these things just as we do who are actors in them, there would be such a howl raised that it would set some of the superannuated old grannies, who undertake to manage these things for the dear people, to thinking.

The fact is that I am getting to regard the navy as a most stupendous & costly humbug—now & then some actors in this expensive play happen by good luck to stumble upon a good thing & they sparkle above the warlike horizon like a rocket. Sensation writers make them a text for numberless items—they receive the thanks of Congress & the credulous public accept them as heroes of the first water—while the real heroes who have done the work & occupy subordinate positions are passed by unnoticed.

William F. Keeler to Anna Keeler, June 14, 1862

Standard histories of the *Monitor* proceed directly from the battle and its immediate aftermath to the ship's loss in a gale. Nearly every book on the topic follows this narrative form, balancing Hampton Roads with the war machine's heroic, violent end, sometimes including the *Virginia's* demise in May 1862.[1] These retellings, however, are highly selective. The Battle of Hampton Roads occurred in early March, and the vessel sank on New Year's Eve, nearly ten months later. The standard accounts elide more than 90 percent of the *Monitor's* operational life. True, the ship never again encountered the intense public drama of its first forty-eight hours at sea, but precisely for this reason the remainder of 1862 formed a critical part of its history. During these months the "revolutionary" new weapon became an operational element of a naval force. Any understanding of the *Monitor's* significance in the history of technology must take this transition into account. In response to criticism that the *Monitor* could not fight at sea, Ericsson often claimed it to be suited primarily for harbor and river service. The summer of 1862 proved that, at least for riverine operations, that claim was unfounded. During these long, ignominious months, the superweapon, packed with heroes, entered the monotonous, uncertain, and brutal world of routine warfare.

At first the crew found notoriety equivalent to military success, public glory equal to military prowess. Performance neatly established their own heroism. All contingency surrounding the vessel seemed settled, the questions foreclosed. The crew should have enjoyed their good fortune and welcomed the holiday from danger. Yet they had a lingering discontent. Their nemesis still existed, neutralizing their force even as they held it at bay. Only seeking out and destroying the *Virginia* would satisfy the *Monitor* crew.

Once loosed on the public, however, the image of glory had unintended effects. The ironclad gained value as a symbol as well as a weapon; and an emblem of victory could quickly become an emblem of defeat. Unwilling to risk such a defeat or, worse, capture by the enemy, the navy maintained strict control over the *Monitor*'s movements, managing the vessel from the secretary's office and sometimes even directly from the White House. As Keeler wrote in frustration, "I believe the department [of the navy] are going to build a glass case to put us in for fear of harm coming to us."[2] Hampton Roads had occurred only two days out of port, but the *Monitor* had a long deployment still ahead. It spent the spring and summer of 1862 suffering the combined effects of Northern strategy and Southern weather.

Closing Technical Questions

Although questions lingered for the crew, John Ericsson and the navy quickly declared a technical victory. Assistant Secretary Fox's conversion, completed by witnessing the battle, had material consequences. "I am glad you have become a supporter," Ericsson wrote to him on March 15. A few days later he declared Fox "a perfect master of the subject" of monitors, praise he heaped on precious few naval men, even engineers. Ericsson now could finally kill the rival construction projects, an effort he saw as a national duty, "preventing an imperfect thing from being got up at the moment when the strong arm of government directed by skill like your own, is carrying the new system into practice."[3] Closure on the contract soon followed, as the navy made the final payment of $68,750 to the *Monitor* consortium on March 14.[4]

For all the intricacy of naval warfare, its numerous subtle variables and tradeoffs, the combat at Hampton Roads foreclosed the technical debate. The navy declared the *Monitor* design superior and held to the conclusion at least long enough to initiate a major building program. On March 17, just a week after the battle, the navy ordered from Ericsson and his associates six more gunboats "on the plan of the *Monitor*," four to be completed before the end of July and the other two before the end of August. As a typographic symbol of the triumph, the *Monitor* lost its capital letter and became "monitor," a type of ship, these new six the "*Passaic*-class monitors."[5] A few months later the navy contracted with Ericsson for two "big class," monitors, the *Dictator* and the *Puritan*. These huge, potentially seagoing monsters measured nearly twice as long as the original and cost more than a million dollars. Ericsson's victory lasted beyond this immediate reaction: for the remainder of the war, monitors would dominate the Union ironclad programs. Of the eighty-four ironclads subsequently laid down during the war, sixty-four were of monitor or turreted types (though not all were built by Ericsson and his allies).[6] The *Monitor* "experiment," so uncertain at first, became a success, even though the other two experimental vessels, the *Galena* and the *New Ironsides,* had not even left their shipyards.

New contracts and declarations of success, however, did not solve the *Monitor*'s problems, nor did they end the controversy. Much to Ericsson's annoyance, questions about the technical merit of the *Monitor* refused to die. Unlike Ericsson, Assistant Secretary Fox had actually been aboard the Monitor; when he witnessed its triumph, he also witnessed its problems. Once again, they centered on living conditions. Fox offered the inventor some suggestions: make the ventilator intakes higher above the deck, move the galley aft so that it could vent exhaust when trimmed for combat (and allow the crew to eat during battle), and generally improve the comfort of the crew. In April Fox visited again, this time during a gale. The *Monitor,* even though at anchor, "was rather uncomfortable, being pretty well under water. Her deck leaks some [a fact confirmed by the crew] and as the rain cannot be removed, this point should be looked to in others." Ever the astute administrator, Fox realized he needed the support of sailors and officers in order to make the monitors a success: "Accommodations, ventilation, and comforts are indispensable to enable us to make the sea-going vessels popular."[7] Fox also stressed the strategic significance of the crew's well-being: "These low craft must be made perfectly comfortable for all hands in all weather if we wish to succeed in them as regular cruisers, a point I desire to obtain."[8] Usually, Ericsson ridiculed these points as the nostalgia of naval officers used to comfortable, obsolete ships, but he could not reject criticism from the powerful assistant secretary, now his staunchest supporter.

Even this early in the *Monitor*'s career (it had been out of the shipyard for less than a month), practical experience was accumulating. Ericsson's professional peers, engineers Alban Stimers and Isaac Newton, had run the vessel at sea and in battle, and now they had technical suggestions of their own. Newton, for example, modified the ventilation system to make the ship more comfortable. For the new monitors, he recommended installing a separate blower for the air circulation system, because "the blowers in the fire room do not perform this function effectively."[9] Stimers thought the pilothouse should be moved to atop the turret, and Ericsson agreed.

Fox, impressed with this expertise, suggested that Ericsson journey to Hampton Roads and see the condition of the vessel for himself—the knotty problem of habitability was best assessed by personal experience. He could then learn how Stimers's and Newton's experience with the machinery might help improve the vessels the navy had just ordered. Ericsson refused, oddly writing to Stimers as though he had been invited to operate the machinery, not to evaluate it: "What can I do, you all know your business so well that my presence is useless." Ericsson argued that he already had ideas far in advance of the current design. "The *Monitor* was not half completed when I saw what *might* be done, being thus far in advance of the structure that Mr. Stimers has practically tested." Ericsson devalued practical knowledge, even that of engineers, in comparison to his own insights as a designer and constructor. "Time will tell," he wrote with more wish than certainty, "that nothing has been lost by my not visiting the *Monitor* as you suggested."[10] Actually, much was lost; his contempt for experience would long plague the monitors.

Deterrence

Aside from engineering, the new ironclad also had problems with personnel and operations. In the one week after the battle, the *Monitor* had three separate commanders. With Worden wounded and in hospital, First Lieutenant Samuel Dana Greene took command. The 20-year-old Greene lacked the necessary experience for the job, however, and Fox replaced him after only a day with Thomas Oliver Selfridge, Jr. (Greene remained executive officer of the *Monitor* until its sinking). Selfridge had been on the *Cumberland* when the *Virginia* attacked and saw most of his gun crew blown apart by shell fire before the ship sank under him. After only a few days, a more senior officer, William Nicholson Jeffers, replaced Selfridge. An ordnance expert and a protégé of John Dahlgren, Jeffers was probably selected by Fox because he could technically evaluate the capabilities of the vessel. Some evidence even suggests that Jeffers was Fox's first choice as the original commander of the *Monitor*, but that Fox chose Worden because Jeffers was unavailable.[11]

The stiff disciplinarian Jeffers commanded the *Monitor* for more than four months, longer than any other man. He was never popular with the crew, although the officers respected his competence and technical expertise. Keeler disliked him intensely, and the feeling was mutual, reflecting the captain's low opinion of volunteer officers. "With him the navy is the pivot of pretty much all of creation & himself is the very center of that point," complained Keeler. George Geer thought Jeffers was anxious to prove himself equal to Worden, and Geer frequently reported to his wife the entire crew's hatred for the man.[12] Compounding the frustrating changes of command, the *Monitor* now belonged to a larger naval force; it was no longer an autonomous, special weapon. As part of the North Atlantic Blockading Squadron, Jeffers and the *Monitor* now came under the command of Rear Admiral Louis Goldsborough. Stung by the devastation of the *Congress* and the *Cumberland* on March 8 (for which he was embarrassingly absent from Hampton Roads), Goldsborough compensated with extreme caution lest similar events befall the *Monitor*.

The crew itched for a rematch, convinced they could beat the *Virginia* in another encounter. "I see you are highly elated at the success of the *Monitor*," George Geer wrote to his wife on March 15, "but hold on we have only commenced. You will soon hear more from us we expect." A renewed fight seemed imminent: "Our men are wild with excitement," he wrote a few weeks later, "and will be very much put out if we do not go up ther [to Norfolk and ferret out the *Virginia*]."[13] Incendiary shells were brought aboard in hopes of burning the enemy inside his own iron enclosure. These infernal weapons made the crew uncomfortable, however, as they contemplated the possibility of their own fate as the target of fire. "We fear being boarded more than anything else, by large bodies of men in Small boats or fast Steam boats in some thick dark night," Keeler wrote to Anna, nonchalantly advertising the new weapon's weakest point to a noncombatant.[14] Such fears were well founded: the Confederacy had learned of the *Monitor*'s weaknesses and planned to exploit them to capture the vessel. Several months later, the *Monitor* captured a gunboat, the *Teaser*, commanded by Lieutenant Hunter Davidson, who had been aboard the *Virginia* during their fight. The *Teaser* contained Davidson's sketches of the *Monitor* and detailed plans for boarding and capture.[15]

To the crew, the *Monitor*'s next move seemed obvious: it should head to Norfolk, find its foe, and "get the rat in his hole."[16] But despite the showing at Hampton Roads, the naval command did not trust the new ironclad as an offensive weapon. Secretly, its utility remained in question, and its perceived vulnerability exaggerated the nervous caution with which Union leaders, including Lincoln, treated the *Monitor* (the president visited the ailing Captain Worden in the hospital and learned of the vulnerability to boarding). On March 10, the day after the battle, Goldsborough received the following order, through Fox and Secretary Welles: "It is directed by the President that the *Monitor* not be too much exposed; that in no event shall any attempt be made to proceed with her unattended to Norfolk" to attack the *Virginia*.[17] Fox modified the order slightly as he telegraphed back to Washington that the "*Monitor* shall take no risk excepting with the *Virginia*."[18] Goldsborough himself ached for action: "I expect to sink her in ten minutes from the time I put at her," he boasted to his wife. Even so, he held back: "Nor shall I allow myself to be caught in any of her traps. It would gratify her excessively if I could be allured to Newport News by feint or foolish operation on her part." Goldsborough, with the navy's support, confused discretion with courage.[19] The *Monitor* waited.

On April 11 the *Virginia* left the heavily defended Norfolk harbor, followed by a small fleet of gunboats. Keeler, impressed by the theatrical sight, observed, "I tell you she is a formidable looking thing. I had but little idea of her size & apparent strength till now." The paymaster bristled for a fight; this time he was promised a more active role, operating the machinery that turned the turret. Neither side, however, wanted to risk defeat, so the two ironclads merely threatened each other over the course of the day, steaming back and forth and firing fruitlessly at a distance. Three small Union vessels in exposed positions were captured, but "except this, nothing further of any importance occurred," Goldsborough reported to Secretary Welles.[20] Yet, no matter how ineffectual, such a dance could quickly grow deadly. Keeler felt the strain: "The same comedy I suppose will be enacted day after day, for I don't know how long, though how soon it may be turned to tragedy none of us can tell."[21] Even Admiral Goldsborough, himself largely responsible for the stagnation, complained that "nothing is more trying than hanging by the eye-lids."[22]

As the spring became hot, ominous plans did seem to be brewing. On May 4 a black cloud from the *Virginia*'s engines appeared behind Sewall's point, "like the genius of an evil omen." The *Monitor* quickly prepared for action, removing the awnings, lowering the smokestacks, and covering all openings with iron hatches. The prospect of a fight energized the crew. "Everybody seemed to step with a livelier gait. Countenances which had been clouded with discontent for the past six weeks, now fairly radiated with satisfaction." Visitors aboard hurriedly shuffled off, including a woman who nearly fainted. Keeler added a quick postscript to a completed letter and sent it off to Anna with the departing tug, "expecting to be able to date my next from the interior of the 'Big Thing'" (as he called the *Virginia*). Again, disappointment followed expectation. "The beast" slowly returned to Norfolk without incident. Keeler, his appetite for battle whetted, saw only ironic combat with food: "We were ready—so was our dinner, on which we immediately commenced a vigorous attack, carrying everything before us by assault, passing the outer works (soup) without difficulty & carrying the stronghold of roast beef at the point of the fork."[23]

Lack of confidence at the top drifted down to the crew. They came across a Southern newspaper that reported that the Yankees had lost faith in the *Monitor*, as indicated by their choice not to engage the rebel fleet when it presented itself. Despite its enemy source, the criticism stung. "On board the *Monitor*, we begin to feel that like the turtle we would like to draw in our head, go to the bottom and burrow into the mud out of sight."[24] Discipline frayed, the crew drank and fought. Geer lamented their "very easy time," bristled at missing the "fun" of supporting the Peninsula Campaign at Yorktown, and cursed Goldsborough for forcing them to "lay here like an old coward."[25] Keeler complained to Anna about "blockading duty in a diving bell."[26] Ambitious Alban Stimers began to think the *Monitor* would no longer satisfy his aspirations. He wrote to Secretary Fox, "If it appeared to me there was any chance of our having a fight with her [*Virginia*] I should much prefer to remain here where I am and take part in it, but it really looks to me as if I might remain another month without being any nearer the proud distinction of having assisted to destroy the formidable monster than I am now." Stimers recognized that the *Monitor* no longer defined the technical frontier—"I am daily losing the opportunity of influencing the designs of the new Ericsson batteries." He soon left for New York to work on the next-generation machines.[27]

The *Monitor* was charged with supporting the Peninsula Campaign. To the navy that meant keeping the *Virginia* bottled up, but not necessarily defeating it. Similarly, the presence of the *Virginia* kept Northern warships out of the James River and away from Richmond. Ultimately, the two new technologies neutralized each other, serving defensive rather than offensive purposes. Throughout the spring, the crew realized that their offensive value was declining while their display value was growing. Isaac Newton voiced his discontent in a letter to John Ericsson in early May: "We are moored in precisely the same position as which we have been in ever since the combat on the eighth [*sic*] day of March, chained fast by the bonds of red tape and old foggysm."[28] Visitors continually made the easy day trip from Washington to view the vessel, but the crew wearied of showing off. For them, stagnation reopened the question of the supposedly victorious battle; they became increasingly sensitive to accusations that the *Monitor* had prematurely broken off the fight. In an admiring, plaintive letter signed "the Monitor boys," the enlisted crew wrote to the wounded Captain Worden and asked him to return. "The Norfolk papers say we are cowards in the *Monitor*," they wrote. "All we want is a chance to show them . . . for you are [*sic*] Captain We can teach them who is cowards."[29] Keeler, however, conveyed the most subtle understanding of the *Monitor*'s politics: "The fact is the Government is getting to regard the *Monitor* in pretty much the same light as an over careful housewife regards her ancient china set—too valuable to use, too useful to keep as a relic, yet anxious that all shall know what she owns and that she can use it when the occasion demands, though she fears much its beauty may be marred or its usefulness impaired."[30] John Ericsson's plan succeeded more than he intended, for his invention now became all admonishment and no fight. Valuable for its *potential* use as much as for its practical effects, the *Monitor* had become a deterrent.

Death of the *Virginia*

On May 7 the drama with the *Virginia* was repeated one last time. The Confederate ironclad came down again, pausing under Craney Island. "She remained there smoking, reflecting, & ruminating till nearly sunset, when she slowly crawled off nearly concealed in a huge, murky cloud of her own emission, black & repulsive as the perjured hearts of her traitorous crew."[31] No longer, however, did the *Virginia*'s forays seem mere intimidation. The vessel now sought an avenue of escape. "The water hisses & boils with indignation as like some huge slimy reptile she slowly emerges from her loathsome lair with the morning light, vainly seeking with glaring eyes some mode of escape through the meshes of the net which she feels is daily closing her in. Behind her she already hears the hounds of the hunter & before her are the ever watchful guards whom it is certain death to pass."[32] Keeler was correct: the net shrank as a Union force dispatched from Fortress Monroe approached Norfolk. On May 8 the *Monitor* and several other vessels shelled the Confederate batteries at Sewell's point in preparation for the landings. The *Virginia* emerged but hurried back before the *Monitor* could engage.[33] The following night, troops landed at a site selected personally by the president. They headed for Norfolk, where they arrived on the tenth. That night, the *Monitor* crew saw a glow from the Gosport Navy Yard; the rebels had set the yard afire in retreat (much as the federals had done just a year before). Amid the flames, Keeler and his companions saw a bright light and heard a dull report, signals of the demise of their nemesis the *Virginia*, as it burned and blew up, intentionally destroyed to avert capture by the Union.[34]

The destruction of the *Monitor*'s enemy should have been cause for celebration. Instead, the crew realized that the questions outstanding about March 9 would never be answered. "The latter information was not so gratifying," Keeler wrote, "as we had ever since the fight looked upon the 'Big Thing' as our exclusive game." Immediately, the *Monitor* steamed into Norfolk and found its heavily armed defenses burned and abandoned. Lincoln and his entourage followed on a steamer. The *Virginia*'s end also ended the *Monitor*'s mission; the ironclad now needed a new purpose.

Up the River

The *Virginia*'s demise not only removed the threat to Hampton Roads and the Peninsula Campaign, but it gave the Union navy free passage up the James River (which had not been an avenue of escape for the *Virginia* because of its deep draft). Goldsborough immediately ordered the *Monitor* and an accompanying flotilla up the James "to reduce all the works of the enemy as they go along . . . and then get to Richmond, all with the least possible delay, and shell the city to a surrender," a campaign he was sure would take two days.[35] Steaming directly into the rebel capital and shelling it into submission (much as they had feared the *Virginia* would do to Washington) seemed the ideal assignment for the invulnerable battery.

After picking up a load of coal, the *Monitor,* with three steamers (the *Aroostook,* the *Port Royal,* and the *Nauguatuck*) and another new ironclad, the *Galena,* formed the James River Flotilla under commander John Rodgers, the *Galena's* captain. This ironclad, the result of Bushnell's original proposal, had been one of the original three recommended for contract by the Ironclad Board. Commissioned on April 21, the *Galena* also rushed into combat, proceeding directly to Hampton Roads without sea trials.[36] Its presence signified that the experimental ironclad program might still have remained open were it not for the events of March 9. The flotilla headed toward Richmond.

Thus ended one phase of the *Monitor's* career and began another, the venture into enemy territory. Keeler marked the change of scene by an observation about women. In Norfolk he observed that "a female could now & then be seen on the verandahs or at the windows of the buildings . . . but not a handkerchief was waved nor a cheer given to welcome us."[37] As the *Monitor* went through Jamestown, slaves welcomed the victors. White men in Jamestown professed to be Union sympathizers, but Keeler noticed they kept referring to "our" army when speaking of the Confederacy and "your" army when speaking of the Union. The atmosphere was tense: "We went ashore well armed & the guns of the fleet were trained on the place." The banks of the river, though beautiful, seemed similarly hostile, and when the men rejoined the ship, guards had to be posted to protect against sharpshooters. Sure enough, shots began to ding off the heavy armor, and "all suspicious clumps of trees & bushes were probed by canister & Schrapnell" from the gunboats.[38] Far from the Union stronghold of Fortress Monroe, the machine now entered a dangerous garden.

Drewry's Bluff

Richmond was not as easily engaged as the *Virginia* had been. The Union force could go only as far as Drewry's Bluff, several miles outside of the city. This turn in the river had been heavily fortified and obstructed with rocks, debris, and sunken ships, "to prevent the ascent of the river by ironclad vessels." On the bluffs ninety feet above the river (later renamed Fort Darling), a number of naval guns defended the point, one even manned by former crewmen from the *Virginia.*[39] Early in the morning on May 15, the Union flotilla attacked; the guns in the earthworks opened fire in response. The Union boats, unable to maneuver in the narrow river, were forced to anchor and wait it out, "a perfect tempest of iron raining upon & around us to say nothing of the rifle balls which pattered on the decks like rain." The guns of the *Monitor* could not elevate enough to hit the batteries on the bluffs, so the ship drew back for a shallower angle of fire. Increased range, however, rendered the guns ineffective against the fortifications. Armor on the *Monitor* (as well as distance) protected the ship: though hit three times, it sustained no damage. Nevertheless, staying belowdecks on the muggy, hot day proved punishment enough: "At times we were filled with powder smoke below, threatening suffocation to us all . . . Some of the hardiest looking men dropped fainting at the guns." Habitability problems had practical consequences in battle. Captain Jeffers reported, "I was obliged to discontinue the action for a quarter of an hour and take the men below to the forward part of the ship for purer air." The ship's log reported, "Several of our crew sick . . . owing to river water and foul air in the ship."[40] (See map, page 106.)

The *Galena* did not fare nearly as well; it attracted the bulk of enemy fire—deadly plunging shot—at Drewry's bluff. "We soon began to see that she was being roughly used," Keeler reported.[41] Unlike the armor of the *Monitor,* the *Galena*'s exotic interlocking plate, designed by John Winslow, did not hold fast. Even those shots that did bounce off dislodged dangerous shards of wood and iron inside the vessel, wounding and killing crew members. Afterward, the *Galena*'s horrified surgeon likened the inside of the ironclad to "a perfect slaughterhouse," with thirteen killed and eleven wounded.[42] "We demonstrated that she is not shot-proof," commander Rodgers reported to Goldsborough with understatement that would be comical were it not so tragic.[43] Keeler had the unfortunate experience of going aboard the *Galena* and seeing for himself the results of failed iron armor—strengthening his appreciation for the *Monitor*'s own, however uncomfortable.

> The scene that I witnessed there is beyond the power of language to describe, to say that she looked like a slaughterhouse would convey but a faint idea of the appearance of her decks, they were literally and without exaggeration a slaughterhouse of human beings . . . twenty five wounded were groaning in agony—the sight was awful, horrible in the extreme, too fearful to look at & impossible to describe.
>
> The shell, though performing its errand with such murderous accuracy would sometimes commit strange & even laughable freaks—the performance of the bull in the crockery shop is nothing to the antics of a shell in the Galena's china closet.
>
> Here was a body with the head, one arm & part of the breast torn off by a bursting shell—another with the top of his head taken off the brains still steaming on the deck, partly across him lay one with both legs taken off at the hips & at a little distance was another completely disemboweled. The sides & ceiling overhead, the ropes & guns were spattered with blood & brains & lumps of flesh while the decks were covered with large pools of half coagulated blood

& strewn with portions of skulls, fragments of shells, arms, legs, hands, pieces of flesh & iron, splinters of wood & broken weapons were mixed in one confused, horrible mass.[44]

The gruesome tale cannot have been comforting to Anna.

After the *Galena's* punishment, the Union flotilla retreated downriver to await reinforcement. The defenses at Drewry's bluff would resist Union advances for the remainder of the war. There stalled the *Monitor,* a powerful, sophisticated war machine in the midst of a hostile river. Keeler, supervising a guard watch through the night, noted, "The idea that we were in the heart of the enemy's country, only ten miles from Richmond, did not add a great deal to the pleasure of my meditations."[45] By May 17 the danger became more immediate, though still unseen. "Not a man could shew himself on the decks without a ball whizzing by him . . . one passed between my legs & another just over Lieut. Greene's head." Confederate sharpshooters, by confining the men below, converted the ironclad into a sweltering prison. Lieutenant Greene later wrote, "Probably no ship was ever devised which was so uncomfortable for her crew, and certainly no sailor ever led a more disagreeable life than we did on the James River, suffocated with heat and bad air if we remained below, and a target for sharp-shooters if we came on deck."[46]

Eventually, the men ventured out to the riverbanks. One afternoon, in a seemingly safe spot near City Point, Virginia, Keeler and several others took a small boat from the *Monitor* and went ashore to have a look around. While they examined the local rail depot, a slave ran up and warned of enemy soldiers approaching. Keeler and his companions began assembling when "a smart fire of rifles was opened on us." The men began rowing to the protection of the *Monitor,* which was moored out in the river. The gunfire continued, killing two officers from the gunboat *Wachusett* and wounding several others; four officers and five enlisted men were taken prisoner. Keeler, though shaken, made it back safely. The warships responded to the attack by shooting into the town and firing shells and canister into a beautiful house on the shore. "An old man & quite a pretty looking young woman (his daughter) came off to us to beg us to spare the place, as they supposed we intended to destroy it & were terribly frightened."[47] Captain Jeffers called them accessories to the murder of his men and forced them to evacuate their home.

City Point and several similar episodes transformed the Northern sailors' anxiety into violence. "It is well understood along the river," wrote Keeler, "that if we are fired upon from the banks, the buildings, if there are any upon the premises from which the firing proceeds, will be shelled."[48] The destructive outbursts reflected the men's frustration at their impotence in opposing attacks from shore. Though designed to fight other ships, the *Monitor* confronted soldiers and civilians defending their homes. Both Keeler and George

Geer (and the *Monitor*'s logbook) described the Southern harassment as "skulking," a term used from the colonial wars to Viet Nam to describe the confrontation of state militaries with local warfare.[49]

Sweltering on the River

Unable to pass Drewry's bluff, the *Monitor* spent late May and June idle, further plagued by heat, rain, and mechanical problems. One of the engine cylinders failed, and engineer Isaac Newton reported that Goldsborough would not even allow him to repair the ventilation system, because the machinery could not be shut down for even an instant, pending a possible fight. It had been running continuously since February 20, he reported to Ericsson, but

> there is still no prospect whatever for us to be engaged in any active duty or to be allowed time to overhaul the machinery. We are doing nothing here but swelter in this river . . . *It appears to me that for some reason or other ComDr. Goldsborough has determined to keep us in this river for some time yet, and when the occasion comes for sending us to sea down the coast or elsewhere, we will be hurried off without a moments notice and without a chance to do any overhauling whatsoever.* I suppose this may annoy you.[50]

Cut off from mail, running low on provisions, the *Monitor* crew's morale suffered. Doing their best to survive in the combination of torrential rain and sweltering heat, when safe from sharpshooters they cooked on deck and took refuge from the sun on the shady side of the turret. A dead slave's body floated by. Much of the crew became ill from drinking bad water from the river. One of the blower belts broke, and the temperature rose to 150° in the galley, 125° on the berth deck where the crew slept (with both blowers working, it was a more comfortable 132° in the galley). Some of the men passed the time gambling and fighting; they read newspapers and learned of McClellan's difficult time on the peninsula. Entries in the ship's log became scrawled and diagonal, in contrast to the orthogonal lines penned with care the previous winter. Geer, jaundiced and suffering from what may have been hepatitis, studied engineering books in hopes of winning a promotion (he was subsequently promoted from fireman to engineer yeoman and later to third assistant engineer).[51]

Keeler, uncomfortable, out of touch with Anna, and lacking entertaining visitors, expressed his frustration through irony. In his letters to Anna, he always put his location in the header, listing "off Fort Monroe," or "tied up in Hampton Roads." Now, however, he labeled them with dark humor, "out of humanity's reach" and "among the Bullfrogs." "Here we lie, day after day & week after week, prisoners to all purposes. Out of humor with ourselves

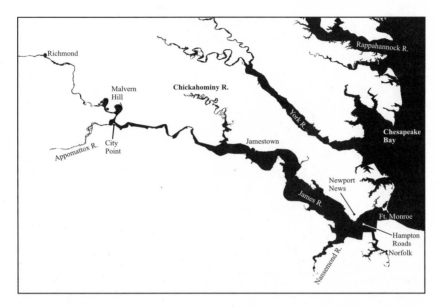

The James River, in Virginia, where the *Monitor* spent the summer of 1862. Note the peninsula between the James and York Rivers, site of McClellan's failed Peninsula Campaign, which the *Monitor* supported.

& the world generally." During June Keeler's exasperated letters became long and monotonous. He wrote to busy himself. "The only fear I have is of getting eaten through by rust."[52] Adding to the irony, the vaunted warship became something of a mockery in these difficult conditions. Forbidden to forage off the land, the crew had to keep its own food aboard, "a portion of our iron deck has been converted into a stock yard, containing, just at present, one homesick lamb, one tough combative old ram, a consumptive calf, one fine lean swine, an antediluvian rooster & his mate."[53] One day, the galley caught fire and threatened the magazine before it was quenched.[54]

As usual, an encounter with women marked the period for Keeler. The *Monitor*'s boat transported to the Southern shore several women who had been caught and convicted as spies in the North. As they passed the odd ship, they taunted the captain.

> There followed a volley from the female tongues—"What a funny looking boat."—"Where can they all sleep?"—"Where do you cook?"—"Wish we could go aboard."—"Wonder where they keep their guns?"—"Where's the holes the MK made in you?"—"Captain was all these men on board in the fight?"
>
> "Can't we come on board?"—"Oh Yes, do let us get aboard."—

"I caused the turret to be revolved into such a position that it should show the port holes and the muzzle of our XI [inch] guns, as well as some of the shot marks on the turret. Then I took a position at the side of the vessel, not with a view to have my picture taken, but to see that very thing was properly done," wrote William Flye *(at right in straw hat)* of posing for this picture in July 1862. Flye to Pierce, March 17, 1886, Pierce collection). Note the dents in the foreground by the gunport caused by the *Virginia*'s shells, and the pilothouse in the background, rebuilt with sloping sides after being destroyed by the shell that wounded Captain Worden. Also note the hatch covers for the deck lights, lying open in front of the pilothouse, and the men standing watch on the turret. Standing to Flye's right is Assistant Engineer Albert Campbell. Courtesy of Library of Congress/Naval Historical Foundation.

"Look in 'Harper' [*Harper's* Magazine] its all pictured out there," Says Capt. J.

"Yes but we want to see how you look inside."[55]

The humorless captain ignored the request.

Lincoln appeared on July 9 for a brief conference before firing General McClellan for his failure to press the campaign more aggressively. Otherwise, minor incidents and adventures broke the monotony. A variety of inventions, including observation balloons, torpedoes, even a submarine, caught the crew's attention. They emphasized the thin line between revolutionary weapons and amusing curiosities. The captured rebel gunboat *Teaser* (itself supporting an observation balloon) yielded plans for boarding the *Monitor.* Keeler carefully read letters from the wife of the *Teaser's* commander and narrated them to Anna. The commodore's injunction against foraging broke down; Keeler went ashore and chased a few pigs escaped from a farm. He and other officers indulged in a bath: "We rowed up a sort of creek some distance from the vessel, found the banks beautifully diversified with pond lilies & coarse rushes & thickly populated with frogs &c&c—disrobed & found the water about knee deep."[56] Some good news came too. Congress abolished the spirit ration for the navy on July 14, making a bright day for Keeler.

In July an "artist," James Gibson, came aboard and took several photographs (in stereo pairs) of the officers and crew. One image shows the enlisted men casually assembled around their cooking fire. In another, men play a game (possibly checkers), one man smokes, and another reads a newspaper. All the men show the effects of exposure. A clear passage through the group reveals the gunport in the turret and the dents caused by the *Virginia's* shells. The officers posed more formally, fully uniformed, in front of the turret. Keeler, notably sunburned in comparison to his January portrait, appears stiff; he alone buttoned his uniform all the way to the top. A troubled Greene stares intently at the camera. Notably absent from the officers' photo, Captain Jeffers appears in a separate image by himself, next to an empty chair. For a fourth image that focuses primarily on the ship itself, William Flye had the turret turned so the artist could capture both the pilothouse and the turret, the latter with its gunports and shell dents. As a group, the pictures depict the tiring, tense summer. Had the camera captured these images in March, the men would have appeared jubilant, perhaps glowing. Now, after a spring of disappointment and a summer of heat, they look hardened, even bitter. As Alan Trachtenberg observed, "The Civil War camera disclosed debris strewn about, weary men, slovenly uniforms,—soldiers not as heroes but as soldiers."[57] Like the *Monitor,* the camera presented a different kind of man at war.

Soon after the photos were taken, Captain Jeffers was transferred back

The officers of the *Monitor* in front of the turret, July 9, 1862. William F. Keeler is standing, second from right, his cloak buttoned up to his neck. The men are sunburned from exposure on the James River (note Keeler's face in comparison with the photograph on page 55). Also note the ready binoculars (and their sack at left): the vessel is in hostile waters and on alert. The two men in front are seated on hatch covers for deck lights. Robinson W. Hands, third assistant engineer, seated in front on the left, and George Frederickson, master's mate, behind him and to the left, both died when the *Monitor* sank. The others are, top row *(left to right):* Albert B. Campbell, second assistant engineer; Mark Sunstrom, third assistant engineer; Keeler; L. Howard Newman, executive officer of the *Galena.* Middle row *(left to right):* Louis N. Stodder, acting master; Frederickson; William Flye, acting lieutenant; Daniel C. Logue, acting assistant surgeon; Samuel Dana Greene, lieutenant. Bottom row *(left to right):* Hands; E. V. Gager, acting master. Captain Jeffers is curiously absent from the image; he was photographed separately, alone (page 104). Courtesy of Library of Congress/Naval Historical Foundation.

north to help build up the ironclad fleet. "We have all been greatly disappointed in our present Captain.," Keeler wrote, due to his "extreme selfishness & his want of energetic action."[58] Yet the new captain, Thomas Stevens (awarded the command for his capture of the *Teaser*), could no more overcome stagnation than his predecessor. For the remainder of the summer, the *Monitor* continued small adventures in support of the Peninsula Campaign, which reached its own frustrating climax. For example, the vessel headed up the Appomattox River in a vain attempt to blow up the Petersburg railroad bridge in conjunction with a "submarine battery," the *Alligator*. When a union gunboat ran aground, the expedition was abandoned.

At the end of June, during the bloody Seven Days' Battle, the new Confederate commander, Robert E. Lee, repulsed the Union attack. On July 1, at the battle of Malvern Hill, McClellan and his army fell back to a fortified position on the river, Harrison's landing (birthplace of President William Henry Harrison). Rodgers, the *Monitor,* and the flotilla supported the land operations and protected river intercourse for the army, as supplies were brought up and troops evacuated. The *Monitor* and several other gunboats attended close by, within earshot but out of sight of much of the fighting. Keeler again bristled at their absence from a clash. "Oh if our gun boats could only plunge their shot into the rebel ranks. It is terrible indeed to be compelled to sit & listen to such fearful sounds & not be able to give assistance."[59] When the *Monitor* did attempt to support operations by lobbing shells into the fray, the ship was quickly waved off because its shells were falling on union troops. Other naval guns were more effective, however, and the James River Flotilla protected McClellan's flank from attack at a critical time.

Sometimes idleness actually enriched Keeler's writing, as he took Anna on verbal tours. A walk through McClellan's camp begins as one leaves the *Monitor* in a small boat: "If you desire to accompany me just step into the captain's gig—take a seat on the cushions &—'Shove off men'—we're off for the camp. See what four stout fellows we have to pull the oars . . . " He shows her the ships at the dock, the hospitals, the nurses, tents full of wounded, the general's own tent, a reconnaissance balloon ascending. Another such tour, a verbal panorama, comprises Keeler's most moving and detailed representation of the war:

> Come up to the top of our turret & stand under the awning with me & take a look ashore.
>
> The bank you see slopes gently down to the water's edge, covered with fine large trees. Over & between their tops, you get a glimpse of the "great house" . . . Scattered along on the shore are some of the wounded & stragglers from the battle field. Here a fallen tree forms a seat for a long line of poor dejected looking fellows, some with ragged blankets drawn over their head & shoulders

as a slight protection to the beating rain. There another group has sought the shelter of that cluster of young pines.

A tall stalwart fellow stands exposed to the rain on the end of the dock moodily leaning on his musket as though he were bidding defiance to both the elements & enemies. Here a sick one lies stretched out on a cast away plank, not even a blanket to shelter him from the rain. Here one limps painfully along through the mud & water by the aid of an old branch, with a wounded foot from which he has discarded the shoe, the suffering member tied up in a bloody, dirty rag, sinks deep in the mud at every step . . .

Take my glass & look along the marshy bank. There they come straggling along singly & in pairs, tracing their weary way through mire & marsh, over rocks & stumps & fallen trees, & through masses of tangled briers & young pines drenched with the water shaken from their foliage & the pouring rain.[60]

The *Monitor*'s armor and its position on the river protected the crew from more than shells and sharpshooters. For all its discomforts, the *Monitor* saved Keeler and his companions from the brutal experience of the war on land, one distinctly at odds with the clean progress the ship represented.

Captain Jeffers posing with dents in the turret armor. Courtesy of the Naval Historical Foundation.

The *Monitor* crew relaxing on deck in July 1862 during their summer on the James River. The temperature inside the vessel was well over 100° Fahrenheit. Despite their casual attitudes, the photographer and crew have composed the picture so that the dents in the turret armor caused by the *Virginia* are clearly visible. Note the man standing watch with a telescope. Courtesy of Library of Congress/Naval Historical Foundation.

"A Most Fetid Atmosphere"

During this summer, the Department of the Navy could have no illusions about the vessel's effectiveness. Captain Jeffers issued a report to the department that criticized the *Monitor*'s capabilities. Jeffers's missive ranks as the most balanced evaluation by those who sailed on the *Monitor,* because he had expertise in naval science but little investment in the ship's ultimate technical success, although he did aim to explain the failure at Drewry's Bluff. He generally confirmed the weak points that were emerging (and confirmed Keeler's accounts of them): control problems for the captain in the pilothouse, the difficulty of accurately turning the turret, the single-point failure of the leather belts that drove the blowers. Jeffers reserved for special indictment the ventilation problems during the summer.

> When the weather was cold it was quite warm below, but no inconvenience was felt other than the impurity of the air passing up through the turret; but with

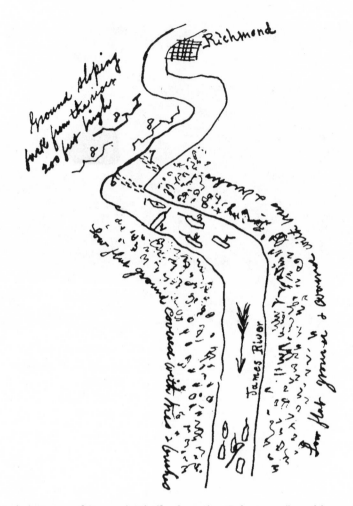

Keeler's image of Drewry's Bluff, where the *Galena* was "roughly used" and the *Monitor* was forced to pull back because it could not fire up at the batteries on the bluff from close range. Keeler's labels include (1) the *Monitor,* (2) the *Galena,* (3–5) the gunboats *Aroostock, Port Royal,* and *Naugatuck,* (6–7) obstructions in the river, (8) rebel batteries, and (9) "Our anchorage of the morning before going into action." From a letter from Keeler to his wife, May 16, 1862. Courtesy of Naval Academy Museum, Annapolis, Md.

the heat of the last ten days, the air stood at 140° in the turret when in action, which, when added to the gases of the gunpowder and smoke, gasses from the fire-room, smoke and heat of the illuminating lamps, and emanations from the large number of persons stationed below, produced a most fetid atmosphere, causing an alarming degree of exhaustion and prostration of the crew.

. . . If the hatches were all closed (as they must be at sea) in this warm weather, the crew would be unable to live for forty-eight hours shut up. Quite one third of the crew are now suffering from debility; there being no shelter on deck, they have to keep below to avoid sharpshooters.[61]

Jeffers saw the bad air as something more than the discomfort Keeler complained of: it was a liability in combat, as Drewry's Bluff had demonstrated. "The vessel cannot go to sea until this defect is remedied." Jeffers gave his recommendations for modifications to the *Monitor,* but he concluded that wooden ships, shell guns, and forts had not been superseded. Most important in Jeffers's report, however, is his written understanding of the politics surrounding the new vessel:

The opportune arrival of this vessel at Hampton Roads, and her success in staying the career of the *Merrimack,* principally by the moral effect of her commander's gallant interposition between that vessel and the *Minnesota,* caused an exaggerated confidence to be entertained by the public in the powers of the *Monitor,* which it was not good policy to check. I, however, feel that I owe it to you, sir, as the commander of the fleet, and to the department, to put on record my deliberate opinion of her powers.[62]

Jeffers left the command soon after making the report, although evidence suggests that his transfer was not due to criticism but to poor health, itself possibly brought on by the vessel's defects. Ericsson called Jeffers's report "a pernicious document" and assured the navy that ventilation problems were already eliminated on the newer monitors.[63] Jeffers nonetheless made an important point: the public image of the machine, its "moral effect," diverged from its technical capabilities, but both could help prosecute the war. He communicated that conclusion to the highest levels of the navy.

Keeler, too, saw the disjuncture between image and experience. In keeping with the navy's desire not to check public confidence in the *Monitor,* no reporters had been allowed up the river and the crew were instructed to give no information about their movements. Reading of the Peninsula Campaign in the newspapers, the crew found that the *Monitor* was still generating good press in the North, despite its stagnation. As Keeler wrote, "The gullible public are duped into believing that we are shelling batteries, making reconnaissances, supporting McClelland . . . they are very little aware that we are having a nice little game of hide & seek."[64] Similarly, George Geer wrote to his wife, "I do not wonder you are worid [worried] at what you read in the paper

... about the doings of the Monitor but they are all bosh we have not had our Anchor up Fired a Gun or been of the least use or servise except to act as a scare crow for most of one month ... so don't believe any thing you hear or Read about the movements of the Monitor until I write you."[65]

When Keeler left the vessel for a few days to get provisions in Norfolk in late July, he found, to his surprise, that the public did not share his cynicism. The *Monitor* still had great currency. "Having been so long on board the *Monitor* she had with me become a common place thing & I supposed had in great measure lost her prestige, but I found it quite otherwise. People seemed to regard her as a sort of irresistible war monster and any one from her as something more than human." Keeler was congratulated, offered free meals, and invited to stay in people's homes. The *Monitor,* he reported, is "open sesame wherever I go" (although two Southern women crossed the street rather than walk by him).[66] If the vessel could not conquer Drewry's Bluff and capture Richmond, it could at least nourish the public and threaten the enemy with its impregnable countenance. After two days in Norfolk, Keeler returned upriver, his taste whetted for the acclaim he and his shipmates could still enjoy. The glimpse of renown made life on the James all the more frustrating; he ached to get back to the cool breezes of Hampton Roads, site of the original triumph. He did not wait long.

One August morning, the crew awoke to find that the entire army, its equipment, and its transports, had disappeared. McClellan had evacuated the peninsula. Remarked Keeler: "Contrabands, wagons, & artillery, all were gone & nothing left to mark the spot till now occupied by McClelland's large army with its immense amount of material, but a smooth, extensive, plain of yellow dust."[67] Geer commented, "I am very much puzzled to know where the Armey went they went off so quick and mysteriously."[68] Feeling now not only defeated but alone, Keeler fell into melancholy. The mood deepened when he received word of the death of Anna's brother, Melzar Dutton, in the Battle of Cedar Mountain in Virginia. Finally, the dreadful summer ended. The Peninsula Campaign terminated, the James River Flotilla disbanded. Its gunboats dispersed and the *Monitor* headed downstream, arriving in Newport News on August 30. The most public weapon of the war had spent nearly four invisible months up the river.

Back to Blockade

Now the *Monitor* repeated its earlier mission. It waited in Hampton Roads for the *"Merrimack II"* (actually the *Richmond*), a new Southern ironclad that had been laid down in Norfolk and then evacuated to Richmond for completion.[69] Anchored to almost the exact spot where its nemesis had destroyed the *Congress* and the *Cumberland,* the *Monitor* lay in wait for the new ship to emerge from the James River, but it never appeared. As before, a number of false alarms sent the crew of the *Monitor* scurrying. The excitement briefly rejuvenated their weary spirits. In general, however, morale quickly improved once they left the confined river. The engineers put down steam for a day so they could repair the machinery. Men were granted shore leave, and fresh food improved their spirits. Geer, who brought a load of peaches aboard, wrote, "I have not been so contented since I have been on the ship"; he suspected the very reason they came down the river was out of concern for the crew's health.[70]

Also in Hampton Roads at this time the crew saw a reminder of the days, which seemed so long before, when the *Monitor* was considered an "experiment." The *New Ironsides,* freshly commissioned, the third of the original ironclads, was tied up nearby, its draft too deep to enter the James. Keeler and the crew went aboard this "huge floating fort," whose majesty stood in stark contrast to the discredited *Galena.* Of the three products of the original Ironclad Board, the European broadside-style *New Ironsides* turned in the best performance, seeing more action than any other union warship.[71] The *New Ironsides* remained a single prototype, however; no more like it were built—the *Monitor* had closed the debate and the monitor type had overtaken nearly all new building.

During these weeks Keeler gradually realized they would never again have a fight as mythical as the one of March 9. If the *Virginia* of that day no longer existed, then neither, in some sense, did the *Monitor* of early March. "Some of us will die off one of these days with *Merrimack*-on-the-brain. The disease is raging furiously, especially among those inclined to old foggyism." Keeler's irony became bitter: "Now we are to live in a state of continued preparation, us 'Monitors' especially, as upon our shoulders rests the Salvation of the whole navy, so you may imagine us close prisoners in the bowels of our iron monster, not a very enviable situation I assure you in the present hot weather."[72] Finally, on September 30, orders came for the vessel to head to the Washington Navy Yard for refitting and repair. The ship's bottom, by this point, was so fouled with marine growth that the *Monitor* required a tow.

On October 2 the *Monitor* arrived in Washington. To prepare the ship for dry dock, the crew emptied it of all their possessions as they prepared for four weeks' leave. Before they left, however, the ship once again became a showcase. "The *Monitor* & her officers are the lions of the day . . . On Saturday the Yard was thrown open & they rushed in by thousands & thousands." Now the glorious ship was flooded with women, and Keeler found himself back in the role he enjoyed most. "There appeared to be a general turn out of the sex in the city, there were women with children & women without children & women—hem—expecting, an extensive display of lower extremities was made going up & down our steep ladders . . . I made a large number of what would no doubt be very pleasant acquaintances if I had the time & disposition to follow them up—as all that I shewed over the vessel gave me their addresses with an invitation to call." A curious scene occurred during this display: "Our decks were covered and our wardroom filled with ladies and on going into my stateroom I found a party of the 'Dear Delightful Creatures' making their toilet before my glass, using my combs and brushes. We couldn't go to any part of the vessel without coming into contact with petticoats."[73] Keeler probably used the term "contact" literally, because the small size of the interior spaces meant that any two persons who crossed paths would rub up against one another. After many months of passing sweaty sailors in the narrow passages, the crew welcomed the well-dressed female visitors, especially as they climbed down ladders. "Suffice it to say," reported one male visitor, "there were some interesting sights not advertised in the bills."[74] Shortly thereafter, the *Monitor* was turned over to the mechanics in the yard. The crew, granted leave, received round-trip train tickets to New York. Keeler traveled to New Haven to meet his beloved Anna. They spent four weeks together.

Throughout its career, the *Monitor* served real military purposes: deterring the *Virginia,* attempting to reach Richmond, supporting the Peninsula Campaign, and a host of smaller missions. Some succeeded, others failed, because of particular circumstances and the strengths and weaknesses of the machine itself. Keeler's discontent echoed a broader malaise in the Union at the time, as McClellan abandoned the Peninsula Campaign and retreated to Hampton Roads. The war, at one point but ten miles from the Confederate capital, dragged on for three more years. Richmond did not fall until 1865.

To charges about the *Monitor*'s lack of seaworthiness, its supporters often responded that the *Monitor* was best suited to harbor and river service, a claim echoed by subsequent observers. On the James River, however, the *Monitor* failed as a river gunboat: its armor was too thick, its guns too powerful, its interior too hot and cramped. Because of the position of the pilothouse, the vessel could not even fire straight ahead, an essential feature for riverine warfare.

The experience and the frustration of the *Monitor* crew on the James River in 1862 derived from these local, military circumstances and also from the general, changing nature of war. Warfare on land, of course, could also move quickly from victory to brutality, from triumph to terror. But it did not bring the same symbolic and technical registers, the incessant contrast between expectation and accomplishment, that mechanical war brought to the James River in 1862. Keeler's experience was not simply that of the plaintive, homesick soldier; he and his comrades were also industrial warriors stifled by a technology's limits. The overwhelming power of the mechanism and its capacity for protection radically diverged from the crew's experience of idleness, frustration, and fear—especially while in the midst of hostile territory. Naval commanders, politicians, the public, and John Ericsson himself confused the luck of a brand new prototype with the success of a mature weapon. Public performance in combat convinced many that the new technology was unquestionably superior, but its users still had much to learn about its strengths, weaknesses, and potentials.

Life on the James had none of the heady risk or glory of Hampton Roads. The crew believed the navy would not use the *Monitor* for combat when they could use it for public relations. Keeler's one-sentence summary, "I believe the department [of the Navy] are going to build a glass case to put us in for fear of harm coming to us" conveys the irony of that situation: the fragile but transparent glass case surrounding impenetrable but opaque iron armor.

Chapter Seven   Utilitarians View the

                                  Monitor's Fight

                                    1862–1865

 *Vague and extravagant notions of the force of the turbulent ele-
ment soon lose their terrors when tested by the unerring standard
furnished by hydrostatic and dynamic laws.*

                                                    John Ericsson, 1863

After a month's leave, Keeler returned to the *Monitor* on November 11, 1862.
In the intervening time, mechanics at the Washington Navy Yard had over-
hauled the vessel. They added new life boats and davits and replaced the smoke-
stack and fresh-air intakes with higher ones to keep out the water. Dents in
the armor were patched over with iron plates and marked "Merrimack" or
"Fort Darling," to denote their origins. The two Dahlgren guns were engraved
with the words "Ericsson," and "Worden." A new awning was installed above
the turret and a new blower engine for the berth deck, and workers added
"many other little conveniences which would have added greatly to our comfort last summer."[1] Benjamin Isherwood performed a series of experiments on
the engines and boilers, precisely measuring their power and efficiency. New
white paint made the officers' room brighter and more pleasant than before.

Death of the Monitor

The *Monitor* proceeded back to Hampton Roads, to monotonous but not
dangerous blockade duty. Refreshed from leave and cheered by the cleanup
and the cool November temperatures, Keeler found these weeks the most
agreeable of his time aboard. He gushed about his rejuvenated home: "I doubt
if you could find better at few hotels on shore."[2] Ironically, Keeler's security
coincided with the greatest danger.

Soon they left for a new mission, headed south. Keeler once again hoped the questions that still dogged the vessel would finally be put to rest: "I wish I could whisper in your ear our destination & plans . . . You will have to nurse your curiosity & patience for a little while, when we hope again to make 'the little *Monitor*' a household word."[3] Keeler's next words to his wife, seven days later, had lost this optimism forever. "I have been through a night of horrors that would have appalled the stoutest heart . . . What the fire of the enemy failed to do, the elements have accomplished."[4]

The iron home, so well refurbished, had turned fatal for its inhabitants. On New Year's Eve 1862, the *Monitor* foundered in a gale off Cape Hatteras, North Carolina; four officers (two of them engineers) and twelve enlisted men drowned. Furious waves pounded the hull as water leaked through the turret and the hawse-hole (the opening that passed the anchor chain out of the hull). The coal fires, soaked from the leaks, could no longer operate the pumps and keep the vessel afloat. Keeler and forty-seven others narrowly escaped to an escort vessel. Several were washed away and others trapped inside. Debates raged about whether the loss was due to improper maintenance, but the ultimate cause of the disaster was the iron raft's basic lack of seaworthiness—Ericsson's vision may have been perfect in the abstract, but it was not properly constructed with the techniques of the day.

To the end, the *Monitor* crew remained ambivalent about their enclosing home. The iron vessel's lack of inherent buoyancy became frighteningly clear as the ocean rushed in, flooding the artificial world. Seaman Frank Butts, as he escaped the doomed vessel, heard the cook screaming at the frightened crew. "He congratulated them for being in a metallic coffin and that the devil would surely pick their bones as no shark could penetrate their graves."[5] Keeler wrote a lengthy account of the night, but for once he found himself unable to describe its true impact. "Words cannot depict the agony of those moments . . . a panorama of horror which time can never efface from my memory."[6]

The loss meant that the questions surrounding the *Monitor* would remain open forever. It had not proved itself the best configuration for new vessels of war. Ericsson had not proved whether the *Monitor*'s defects were artifacts of its hasty construction or fundamental problems. Since the *Monitor*'s status would never be settled by combat, Ericsson redoubled his efforts to settle it by argument. As Hawthorne had predicted, the ensuing controversy called into question the relationship between the technical knowledge of engineering and naval expertise born of experience.

New Monitors, Old Questions

In late 1862 when the *Monitor* sank, the new monitors were just coming out of their shipyards. Ericsson had hoped to refine his design and construction so these would prove the success of the original. They had dramas of their own, however, and did not settle matters. Granted, a number of important problems were fixed on the new monitors: they had better blowers, hulls with conventional lines, and pilothouses atop their turrets. These so-called *Passaic*-class monitors also mounted a monster fifteen-inch Dahlgren gun, in addition to an eleven-incher like those on the *Monitor*. The improvements, by eliminating complications, served to focus debate on the basic features of the low-freeboard, turreted ironclad.

One month before the *Monitor*'s sinking, the second monitor, the *Passaic*, was commissioned. It had a relatively easy passage to Hampton Roads, suffering only a burst boiler and failed steering gear. Yet a reporter from *Harper's Weekly* accompanied the crew on the trip and found it unsettling. "Every wave broke over our low decks," he recounted, "and, like a huge sea-monster, the ship plunged through them, dripping and leaking in a manner unpleasantly suggestive."[7] On New Year's Eve the *Passaic* barely survived the same storm that claimed the *Monitor*. The *Passaic*'s captain, Percival Drayton, who had overseen the vessel's construction in New York and commanded its passage to Hampton Roads, formed a low opinion of the contrivance. The New Year's Eve storm did not improve his evaluation. "My crew, as you may suppose, have been very much overtasked, and could scarcely have stood the fatigue they were necessarily subject to much longer, and were fast breaking down from want of sleep and impure air."[8] *Harper's* dramatically recounted how water threatened to overcome the vessel, how the crew desperately bailed it out, and how the engineers calmly struggled to keep the pumps going. Drayton's chief engineer, George Bright, reported that the soaked *Passaic* had come perilously close to sharing the *Monitor*'s fate.[9]

After the storm, Captain Drayton wrote to Welles, "I think it will be a cause of disappointment if the vessels of this class are looked upon as more than steam-batteries, to be towed from point to point in fine weather."[10] The captain reiterated his position that the overhangs, where the upper raft extended beyond the lower hull, caused hideous noise and damaging shock when they slammed down on the water in heavy weather (even though they were smaller on the *Passaic* than on the *Monitor*). This problem, he wrote, combined with the leaky turret, "will prevent these vessels from ever being safe at sea."[11] Welles took Drayton's criticisms seriously. "The fate of this vessel affects me in other respects," he wrote on hearing of the *Monitor*'s loss. "She is a primary representative of a class identified with my administration of the Navy."[12] The secretary forwarded Drayton's comments to Ericsson soon after the loss of the *Monitor*: "These reports require your very careful and serious attention and I invite your special attention to the weaknesses developed in the junction of the two vessels caused by the very great overhang."[13]

Ericsson gave little merit to Drayton's comments, believing the captain had improperly trimmed the vessel by putting too much coal aboard. Even as his own *Monitor* was sinking, Ericsson called Drayton "utterly incompetent to decide on the merits of this vessel." He continued, "I am more and more surprised at the course of this officer who seems bent on prejudicing everybody against the vessel under his command."[14] His own convictions notwithstanding, news of the *Monitor*'s foundering put Ericsson on the defensive. "I freely admit that I was quite thrown off my guard and therefore omitted to give that strength to the structure which recent experience has shown to be indispensable," he wrote of the problem with the overhangs; but he said he had nonetheless corrected it in the new versions:

> The action of the sea under the extreme ends of our Iron Clads, which *experienced* Admirals and Gallant captains look upon as an insurmountable difficulty, dwindles down to something not at all beyond computation when subject to the investigation of the *experienced* engineer. Vague and extravagant notions of the force of the turbulent element soon lose their terrors when tested by the unerring standard furnished by hydrostatic and dynamic laws. When told of the fearful beating of the *Passaic*'s projecting bow against the sea and the angry surge that follows, the reflecting engineer, so far from being disposed to joining Captain Drayton in his advise to give up the principle, calmly sets to work and estimates the actual force expected and the amount of resistance needed to meet it . . . he proceeds to calculate.[15]

After the *Passaic*'s initial voyage, further operations still failed to resolve questions that the *Monitor* had so dramatically opened. The *Nahant* was commissioned just two days before the *Monitor*'s loss, and, despite the concerns, the *Weehawken* a few weeks later. They too met a gale on their way to Hampton Roads. The *Nahant* retreated to safety, but Captain John Rodgers (who had commanded the tragic *Galena* during joint operations with the *Monitor* the previous summer), took the *Weehawken* through the storm. The feat earned him praise from numerous ironclad supporters, including Secretary Welles, and permanent admiration from Ericsson. "When the Monitor went down the 'I told you so' people were to be met everywhere," a jubilant Gustavus Fox wrote to Rodgers. "You have disposed of them! I congratulate you, your brave act has been of more use to us than a victory. Send an official report of the most minute character, it will be historical."[16] Simply surviving in a monitor was now deemed a feat of heroism, equivalent to military victory.

Further combat experience proved to be equally uncertain. In April of 1863, under pressure from Ericsson and his allies (sometimes referred to as "The Monitor Lobby") to test the new ironclad fleet, Welles ordered a fleet of seven monitors and several other ironclads to take Charleston Harbor, the Confederacy's most heavily defended port. Rear Admiral Samuel Francis DuPont, who was to command the expedition, advised against it but could not overcome Welles's and Fox's faith in the monitors' invulnerability. On April 7, 1863, led by a reluctant DuPont, the attack failed. Recriminations in the aftermath centered on the question whether to blame the commander's performance or the weaknesses of the machines. DuPont was replaced, but not before he brought engineer Alban Stimers before a court-martial. Stimers had been at Charleston as a technical consultant on the monitors, and DuPont brought him up on dubious charges that Stimers had criticized DuPont's command in the newspapers (Stimers was later acquitted). DuPont's successor was Rear Admiral John Dahlgren, the ordnance expert now turned line officer. Although much more the technical expert than DuPont, Dahlgren too failed to take Charleston in similar attempts in the summer of 1863. He reported that the monitors suffered more casualties from heat exhaustion than from any other factor.[17]

In June 1863, the *Weehawken,* still under John Rodgers, defeated and captured the Confederate ironclad ram *Atlanta* in a close-in duel, the most notable success of a monitor after Hampton Roads. Victory was short-lived, however; in December of 1863, a sudden rush of water into the hull sank the *Weehawken* at its mooring, drowning more than 30 men (the 4 officers lost were all engineers in the engine room). Similarly, in the battle of Mobile Bay in August 1864 (the occasion of David Farragut's famous "Damn the Torpedoes" command), the monitor *Tecumseh* sank in thirty seconds after striking a Confederate torpedo, killing 93 of its crew of 114, including 42 of the 43 men in the engine room.[18]

A detailed assessment of the monitor class in the Civil War is outside the scope of this book (indeed, the topic awaits a modern scholar). The important point here is that the monitors' service record in the Civil War, equivocal at best, did nothing to resolve the questions raised by the original *Monitor.* Lively debate continued for years, both in the popular press and in official reports and correspondence. Disagreement centered on both the performance of the *Monitor* at Hampton Roads and the general suitability of the monitor type for naval operations. Ventilation, habitability, and seagoing qualities remained integral to discussions of offensive power, defense, and combat effectiveness. Living conditions—vaguely defined but critically important—stressed the fissure between engineering and experience, between designers and users, to its breaking point.

Ericsson and the Officers: Engineering versus Experience

Much of this debate occurred in the *Army and Navy Journal,* William Conant Church's military gazette of "unquestionable loyalty." It began publication in 1863 in response to the New York City draft riots and the perceived loss of faith in the military. Church admired Ericsson and took much of his own opinion on naval affairs from the inventor. Ericsson frequently published articles in the journal, whose readers included numerous military and civilian officials.[19]

The first issue of the *Army and Navy Journal* printed a letter that criticized the living arrangements aboard the new vessels:

> There must still remain objections to armored vessels which should exempt their opponents from the charge of unreasonable old fogyism. While the protection the Monitors afford is so complete that their officers and crew may escape the perils of battle, they are still forced to accept the alternative of the daily discomforts of a life in the confined and necessarily ill-ventilated apartments of a submerged vessel. If they escape untouched in life and limb, it is at the expense often of constitutions so shattered that they must bear through years of ill health and suffering the pain otherwise concentrated into a brief period of agony.[20]

Signed "old fogy" (old fogies and "fogyism" appear as recurrent themes in the *Monitor* story), this letter came from the pen of *Monitor* engineer Isaac Newton, a man not easily dismissed as ignorant or untutored with the machinery. He had served six months as chief engineer of the *Monitor* and actually became a frequent defender of the monitors' fighting qualities, as well as one of Ericsson's very few close friends in later years. Another letter, from anonymous respondent "Young America," criticized the officers of the old wooden navy for retaining a taste for comfort over combat: "Our seamen, however, have a nobler object in view than mere personal convenience and comfort. Their ambition is fearlessly to do their duty . . . Believe me, "Old Fogy," your day of canvas, wood, pop-guns and comfort is over."[21] The debate continued back and forth for months. "Sailor" wrote in, calling ironclads "perfect coffins for their occupants, even when on a peaceful cruise." Now using his own name, Newton wrote to defend the vessels against the attack. Ericsson, under the pseudonym "Novelty" (the name of his first locomotive), accused commanders of blaming their tools for their own failures of judgment in combat. He admitted that the vessels were unhealthy but claimed to have fixed the problem in the new generations of monitors.

Official correspondence took the issue no less seriously. Here the line officers, who had used the machines in the field, equivocated. After the first failed attack on Charleston, a group of monitor captains (including Drayton and Rodgers) criticized the vessels in a common letter. "We agree that the ventilation may be improved," they wrote. "At present the air from the berth deck which has been breathed by men packed upon it and filled with their exhalations—air from the water closets and from the galley is taken into the blowers and diluted with fresh air and redistributed for use again." They placed blame for DuPont's failure at Charleston Harbor squarely at the feet of the monitors.[22] In 1864 the Joint Committee on the Conduct of the War, put together by congressional Republicans to expose fraud and abuse, turned their attention to the monitors and targeted Gideon Welles. Welles took the offensive, however, and collected reports from his top officers that he published in an edited volume. Included was Jeffers's report from the summer of 1862, concluding that "protecting the guns and gunners does not, except in special cases, compensate for the greatly diminished quantity of fire artillery, slow speed, and inferior accuracy of fire" and arguing that monitors did not obsolete wooden ships.[23] Welles pleaded with John Dahlgren to submit a favorable report, but Dahlgren did not believe that Hampton Roads had firmly settled the ironclad question and he pointed out problems with the vessels. Having commanded his own unsuccessful monitor operation at Charleston, he still concluded that these problems "are susceptible of being remedied partially or entirely."[24] Rear Admiral Goldsborough judged that the monitors certainly had impressive powers of offense and defense in smooth, protected waters, but nonetheless he wrote:

> Their absolute worth, as in these particulars, I cannot regard as entitled to the extravagant merit claimed for it, induced, I apprehend, in great measure by conclusions from the encounters of the first *Monitor* and *Weehawken* with the *Merrimack* and *Atlanta,* without a sufficient knowledge of the facts attending them. That the charm of novelty and construction, or quaintness in appearance, had anything to do with the matter I will not undertake to assert, although I may, perhaps be allowed to indulge suspicion as to probable effect. Popular opinion is not always right on such subjects, nor do I know that it is apt to be when it runs counter to popular naval opinion. At any rate, I do know that the latter is not likely to be very wrong in relation to professional matters of the kind.[25]

Goldsborough echoed Jeffers's earlier opinion that public perception added to the monitors' success and that the weight of judgment ought to rest on the shoulders of naval professionals (indeed, he was the addressee of Jeffers's report).

These line officers were primarily concerned with naval questions of tactics and strategy. None denied that the crews were well protected—monitors had endured hundreds of shots during the war with small injury to the crews—but questions remained about the adequacy of their firepower and whether sacrifices made in speed, comfort, and number of guns were worth the armor's protection. Nearly all agreed on the monitors' unseaworthiness, but most thought it a local problem that could be remedied by design changes. But even if the monitors could be made capable of crossing the ocean, would this be the best configuration for an iron warship? The more traditional *New Ironsides* earned a solid reputation during the war, and its success argued for a scheme more similar to European designs. Furthermore, new large ordnance (such as the new fifteen-inch Dahlgrens and rifled shell guns), developed partly at Ericsson's urging, threatened to penetrate even the best-protected turret, calling into question the value of armor at all. "By 1865," wrote historian Bernard Brodie, "the principle of armor on the warship was definitely on the defensive."[26]

Ericsson, for his part, tirelessly countered the criticism, generating an impressive array of articles and letters that rebutted skeptics with engineering arguments.[27] For him, the issues came down to expertise. "It has often given me pain," he wrote, "to think that our fighting *machines* are entrusted to officers who know nothing of mechanics, and *therefore* have no confidence in their vessels."[28] For example, Jeffers reported to Welles from experience that the guns could not be fired within thirty degrees on either side of the bow: "I tried this experiment myself, and the pain and stupefaction caused by the blast of the guns satisfied me that half a dozen similar discharges would render me insensible."[29] Ericsson, ignoring the captain's stupefaction, responded with trigonometry, "it will be found that by turning the turret through an angle of only *six degrees* from the centerline of the vessel, the shot will clear the pilot house, a structure too substantial to suffer from the mere aërial current produced by the flight of the shot."[30] In the same vein, commanders Rodgers and Worden noted that the monitors were subjected to severe strain in all but the lightest seas. Ericsson replied that "the absence of buoyancy in heavy sea . . . is actually a favorable feature." Traditional surface vessels are pounded by heavy seas, he explained, but "under similar circumstances the Monitor craft becomes partially immersed by the waves which *pass over its decks,* instead of violently tossing it up and down during their oscillations," reaffirming his vision of the monitors as submarine vessels. The captains were confused about the strain, he continued, because of the loud noise the sea made when it lashed against the hull. "An observer, accustomed only to the light, dull sound of a wooden vessel, is startled by the sharp, harsh ring of the metallic hull, and imagines a severe strain where, in fact, nothing but a very natural and harmless sound occurs." Similarly, problems with leaks in the deck openings (from the turret, the hatches, and the skylights) were due to "disregard of my instructions, and the adoption of the sailor expedient" for sealing.[31] The commanders reported their monitors underpowered to make headway in a gale, even under tow. Ericsson responded that such complaints "will surprise all naval engineers who are aware of the engine power applied; nautical science teaches the fact that submerged bodies are but little affected by the violence of a gale."[32]

Any criticism, in Ericsson's view, could stem only from the ignorance of naval officers, from modifications of his plans for construction, or from shoddy construction techniques. That position forced him to break with Thomas

Rowland, whose firm built the *Monitor* hull and whom Ericsson blamed for the construction problems and the weak overhangs. Rowland, he argued, had become "both negligent and overbearing by the money he has made from my undertakings."[33] But Ericsson's case was not bolstered when, in 1864, an entire fleet of twenty "light-draft" monitors, built under the supervision of the ambitious (and dubiously competent) Alban Stimers, were barely able to float. Ericsson legitimately disclaimed responsibility because "the leading principle has been frittered away by changes" made by Stimers, but the embarrassment stung nonetheless.[34]

Ericsson's writings ring with the force of a man absolutely convinced he is right, willing to counter any argument with technical detail—as though by pure will he had brought the *Monitor* into being and by pure will he would ensure a legacy of success. One senses that he and his opponents were discussing different matters entirely, talking past each other. Ericsson's friend William Conant Church, writing in his *Army and Navy Journal,* fairly summarized the debates as follows:

> Captain Ericsson asserts that his ships are seaworthy; his opponents say that two of them have sunk. He replies that the dictator class will not sink; they answer that these have not been tried. They charge that the Monitors lack speed; he replies that when their bottoms are clean they will steam eight statute miles per hour . . . He asserts that they are well ventilated; they reply that in action and in heavy weather they are so close that the men and officers suffer.

As Church put it, "Captain Ericsson is describing a theoretical Monitor, the others are considering its practical use and availability."[35] Naval line officers made detailed critiques based on the actual vessels; Ericsson constantly argued for the platonic *Monitor,* dispelling problems by attributing them to practicalities.

These differences ultimately derived from the tensions between engineering and experience that surrounded the *Monitor.* The officers, Ericsson claimed, did not know how the machines worked, so they could not comment on their suitability. When Drayton criticized the leaking hull, Ericsson replied that his reports "show how necessary it is to receive with caution the statements made and inferences drawn, even by experienced and impartial seamen, in relation to our new system."[36] Arguments born of experience were to be countered by engineering logic. "I should rather trust the judgment of a skilful practical engineer as to the real damage done, than to the opinion of the gallant commanders of these vessels, most of whom know nothing of mechanical matters."[37] The navy's line officers did, however, have one thing Ericsson lacked. They might not have been engineers, they might not have been competent, they might even have been fools, but all of these men had direct experience with the vessels. John Ericsson had never been to sea on a monitor, he had never lived on one, and he had certainly never entered combat in one. John Ericsson, in fact, had no experience with war at sea at all.

Ericsson's unwillingness to credit the officers' criticisms stemmed from the very totality of his program. His ultimate vision, though rarely articulated, was not simply one of ironclad ships, nor even one of new inventions. He dreamed of an entirely mechanical warfare. He believed that naval men misunderstood the monitors because they were not ships at all, "I beg earnestly . . . to call their attention to the fact that they have entered on a new era, and that they are handling not *ships,* but floating fighting *machines,* and that, however eminent their seamanship, they cannot afford to disregard the advice of the engineer."[38]

From his very first treatise in 1854, "New System of Naval Attack," to his wartime correspondence, Ericsson repeatedly used the term "system," to describe the monitors and his other military inventions. This term, in common usage today in military circles ("weapons systems") was, in Ericsson's day, more commonly applied to systems of philosophy than to collections of machinery. Indeed, Ericsson did consider the *Monitor* to be part of a new mechanical philosophy of warfare. In addition to building monitors, Ericsson conjured propellers, torpedoes, torpedo boats, machines for harbor defense, and other military mechanisms. He envisioned the entire endeavor being turned over to designers and handlers of machinery. Combat would then become so terrible that, in the now-tired cliché, war would be made obsolete. In a rare public appearance soon after Hampton Roads, Ericsson praised not the crew but Alban Stimers, the engineer, as the true hero of the fight.[39] "If you apply our mechanical resources to the fullest extent," Ericsson wrote to President Lincoln, "you can destroy the enemy without enlisting another man."[40] The engineer's calculations, Ericsson believed, would ultimately replace the warrior's gallantry, and hopefully his suffering. Ericsson was, of course, not the first to envision a warfare ruled by machinery. Robert Fulton, for example, had similar dreams a generation earlier, as had many others throughout history.[41] In 1862, however, Ericsson could draw on an industrial base well beyond that available to Fulton and similar visionaries, and also on a cadre of expert mechanics and engineers to realize his dream.

Nonetheless, when pressing for the realization of his system, Ericsson failed to recognize that it would depend on thousands of trifles. They would be kept in order only by the skills of the workers who built the machines and the officers and sailors who fought with them. In order to create a successful, sustaining technology, he had to create the users as well as the machines; he had to educate them in living, sailing, and fighting in his new mode of warfare. This man, who created a stunning new ship by issuing drawings from his office, who gathered allies to push his plan through, who saw his brainchild become the hero of the Union, overestimated the power of mind in a technological world, where success still depended as much on details of maintenance and operation as on brilliant design. Monitor captain John Rodgers put it best when he wrote of Ericsson: "He is a genius; and an obstinate fool—He sees what other men do not, and cannot see plain things—he is a genius to be used, not a man of sense to be followed—and yet so cranky and opinionated that [any] doubt at his conclusions is an insult, or a proof of enmity, or gross stupidity unworthy of a thought."[42] To the end, Ericsson insisted that naval officers resisted his new technology because they held tightly to the traditions of their sailing past. Yet Ericsson's opponents were not superannuated old sailing men—he converted men like Joseph Smith and Gustavus Fox fairly easily; even Admiral DuPont opposed the monitors for reasons tied up with his own failures at Charleston. No, Ericsson fought the new naval officers, men with special forms of knowledge in addition to their naval professionalism. These were the men Ericsson failed to convert to his cause: Benjamin Franklin Isherwood, John Dahlgren, Percival Drayton, William Jeffers. They were fighting their own battles within the navy for recognition of mechanical expertise, technical knowledge, and scientific attainment. Dahlgren, for example, was considered a courtier because Lincoln promoted him to admiral on the basis of his ordnance innovations. Isherwood spent his years as engineer-in-chief locked in struggles with powerful private constructors. Of the steam-generation officers, some remained opponents, some retained a cautious optimism, but none became avid champions of the monitors. For these men, a solid alliance with a famous inventor could only have strengthened their cause, but only if his new machine could meet their technical expectations.

Decades after the Civil War, as naval arms races in Europe accelerated amid rapidly evolving ship and ordnance technology, officers, technical experts, legislators, and private builders would join forces on a significant scale. William McNeill called this significant new configuration of state-technology relations "command technology;" its ramifications reach well into the twentieth century.[43] During the American Civil War, however, that mode, though budding, remained in its infancy—hence we must avoid applying modern ideas such as the "military industrial complex" to this period. Ericsson did not make allies of expert officers; his monitor coalition did not include line commanders; he achieved no self-reinforcing cycle of technical development. At the same time, his relationship to the navy involved subtler dynamics than simply the "resistance to change" found in the standard revolutionary accounts. At issue was a new kind of warfare and who would fight it.

Chapter Eight

Melville and
the Mechanic's War

Plain be the phrase, yet apt the verse,
  More ponderous than nimble;
For since grimed War here laid aside
His Orient pomp, 'twould ill befit
    Overmuch to ply
  The rhyme's barbaric cymbal.

Hail to victory without the gaud
  Of glory; zeal that needs no fans
Of banners; plain mechanic power
Plied cogently in War now placed—
    Where war belongs—
  Among the trades and artisans.

Yet this was battle, and intense—
  Beyond the strife of fleets heroic;
Deadlier, closer, calm 'mid storm;
No passion; all went on by crank,
    Pivot, and screw,
  And calculations of caloric.

Needless to dwell; the story's known.
  The ringing of those plates on plates
Still ringeth round the world—
The clangor of that blacksmiths' fray
    The anvil-din
  Resounds this message from the Fates:

> *War shall yet be, and to the end;*
>   *But war-paint shows the streaks of weather;*
> *War shall yet be, but warriors*
> *Are now but operatives; War's made*
>   *Less grand than Peace*
>   *And a singe runs through lace and feather.*
>
>   Herman Melville, "A Utilitarian View of the
>   *Monitor*'s Fight," *Battle Pieces and Aspects of the War*

In John Ericsson's vision of the future, uncanny machinery replaced terrible passion. Hawthorne saw similar changes reflected in the *Monitor*'s iron.[1] In this brave, violent world, cunning contrivances would overtake skill and manhood as determinants of success in war. Herman Melville's poems about the *Monitor* echo these predictions but go deeper than Hawthorne's sardonic musing to expand the *Monitor* crew's own ideas. Biographer Newton Arvin called Melville "the first poet in English to realize the meaning of modern technological warfare, and to render it, grimly and unromantically, in his work."[2] By tapping the *Monitor*'s registers of skill, depth, and irony, Melville's rendering of the ironclad links it to the subsequent history of mechanical warfare.

Melville's Intimate Borrowing

Battle Pieces and Aspects of the War, which contains Melville's ironclad poems, appeared in 1866, his first major publication in nearly ten years. Selling only a few hundred copies, *Battle Pieces* did not replicate the commercial success of Melville's earlier career. He had burst onto the American literary scene in 1846 with *Typee,* his barely fictional and extremely popular account of adventures in the south seas, and had finished his masterpiece, *Moby-Dick,* in 1851. His most recent major work, *The Confidence Man,* had appeared in 1857 after an intense burst of activity: for a period in the 1850s, he averaged about a novel a year. Now Melville turned to poetry, transforming the Civil War into an allegory of good and evil while conveying its intimate, sad experience. *Battle Pieces* consists of seventy-one poems, on subjects ranging from "The Portent," about John Brown, to "The Surrender at Appomattox." Some poems address specific battles, others the psychological and mythical aspects of military experience. Throughout, the "Iron Dome" of the capitol in Washington (still under construction during the Civil War) serves as an imposing figure to tie the poems together. It also evokes the iron enclosure of the *Monitor*'s turret.

In contrast to Hawthorne's "disembodied intelligence" on the war, Melville molded his work from research. He regularly read *Harper's* and several newspapers.[3] He drew particularly on the *Rebellion Record,* a periodical that revealed the war's details more deeply than the newspapers, through diaries, documents and stories, and poems. Melville borrowed from his reading, often turning specific newspaper articles directly into poems, adding metaphors and imagery to achieve a broader purpose.[4] Of the incidents in *Battle Pieces,* none gains more attention than Hampton Roads. Four poems out of the seventy-one engage the events of March 8 and 9, whereas only four others, "Donelson," "The Stone Fleet," "Commemorative of a Naval Victory," and "Running the Batteries," deal with naval themes at all.

The ironclad poems borrow language and imagery from Hawthorne's "Chiefly about War Matters."[5] The two lesser poems, "The *Cumberland*" and "The *Temeraire,*" mirror the standard public vision of the ironclads' impact, the passing of the "wooden walls," and the nostalgia for the days of "navies old and oaken," a time in which Melville himself was at home.[6] The latter poem was, in Melville's words, "supposed to have been suggested to an Englishman of the old order by the fight of the Monitor and Merrimack." It alludes to a painting by the English artist Victor Turner, in which a diminutive steam-tug tows a wooden man-of-war. Now

> The rivets clinch the iron-clads
>> Men learn a deadlier lore . . .
> O, the navies old and oaken
>> O, the Temeraire no more![7]

Two other poems, "In the Turret" and "A Utilitarian View of the *Monitor's* Fight," deal specifically with human and social experience on the new warship. The first takes as its subject one man coming to terms with the iron enclosure. Here the language practically copies Hawthorne's observation of the *Monitor,* that "billows dash over what seems her deck." Melville wrote,

> Escaped the gale of outer ocean—
>> Cribbed in a craft which like a log
> Was washed by every billow's motion.

The *Monitor's* captain, "sealed as in a diving-bell," encounters a spirit, "forewarning / And all-deriding" that chides him for thinking he can protect himself: "Man, darest thou—desperate, unappalled—/ Be first to lock thee in the armored tower?" The spirit warns that what shields him from the enemy may also prevent him from escaping death:

> This plot-work, planned to be the foeman's terror,
>> To thee may prove a goblin-snare;
> Its very strength and cunning—monstrous error!

The captain summons his courage and fights on with the knowledge that his armor may also be his coffin: "Stand up, thy heart; be strong; what matter / If here thou seest thy welded tomb?"[8]

Before most of the crew's accounts were published, Melville saw the terrible potential of enclosure, the possibility of the ironclad's becoming a "welded tomb." His depiction of the ship's psychological effect echoes the fears and ambivalence of those aboard the *Monitor*. In Melville's writing on the ironclad, men struggle with the experience of machinery, trying to come to terms with the complexity of new inventions. He recognized their anxiety and identified its source in an allegory of human engagement with technology. Protection can quickly turn to entrapment, security to destruction, accomplishment to hubris.

Historian Stanton Garner describes Melville's fourth ironclad poem, "A Utilitarian View of the *Monitor*'s Fight," as "perilously close to being the finest poem of the Civil War."[9] Before we consider this poem's relation to the *Monitor*'s story, let us look briefly at the experiences and ideas that led Melville to contemplate the new industrial weapon.

Unlike Hawthorne, Melville never visited the *Monitor* at Hampton Roads (although he may have visited the area later in the war), but he had crossed paths with the vessel and its history.[10] His cousin, Guert Gansevoort, a career naval officer, served at the navy yard in Brooklyn during the early years of the war. Melville visited Gansevoort there in June of 1861, witnessing the frenzied efforts to bring the force of a military industrial facility to bear on making war. More important, Melville may have visited the yard again the following February. By that time the *Monitor* was there, having been just turned over to the navy for outfitting. Gansevoort, now acting commandant of the yard, corresponded with John Ericsson while procuring the two Dahlgren guns for the *Monitor* and may have thought his cousin would be interested in the new device.[11] Melville may have seen or inspected the *Monitor* while it was in port, since he was in New York during the winter of 1862 and left many of his days unrecorded. He may even have attended the launching of the sloop *Adirondack,* the event that so thrilled Keeler during his first weeks in the navy. Gansevoort had supervised its construction and was to command the vessel (it sank under him after grounding the following summer).[12]

Melville also had a profound connection to Nathaniel Hawthorne. That *Battle Pieces* drew on the latter's *Atlantic Monthly* article was no coincidence. Among Melville's myriad influences, Hawthorne was extraordinary; his work, wrote Melville, "dropped germinous seeds into my soul." Melville believed that Hawthorne, fifteen years his senior, had transformed the potential of American literature. The two men met in the summer of 1850 in western Massachusetts and struck up an unusual and brief friendship, corresponding and conversing as Melville completed *Moby-Dick*. In a radiant essay review of that year, "Hawthorne and His Mosses," Melville lovingly praised Hawthorne: "full of such manifold, strange and diffusive beauties." Most appealing to Melville, however, was Hawthorne's disquiet: "You may be witched by his sunlight,—transported by the bright gildings in the skies he builds over you;—but there is the blackness of darkness beyond; and even his bright gildings but fringe, and play upon the edges of thunder-clouds."[13] Melville gave Hawthorne the first copy of his new work, *Moby-Dick; or The Whale,* dedicated to Hawthorne "in token of my admiration for his genius."[14] Hawthorne gave Melville the shocking, allegorical brooding that animates both *Moby-Dick* and his musings on the *Monitor.*

Like Hawthorne, Melville was not new to technological subjects. In the short story "The Bell Tower," for example, a "mechanician," a "practical materialist," creates a Frankenstein-like golem to strike the bell in his babel-like tower. Melville often used machinery as a metaphor and an indicator of social inequality. In "The Paradise of Bachelors and the Tartarus of Maids," wealthy men sit comfortably in a London club, juxtaposed with young women feeding hungry machines in a Massachusetts paper mill. The mechanism steals their vitality as it produces the blank white substrate of writing. "Machinery—that vaunted slave of humanity—here stood menially served by human beings, who served mutely and cringingly as the slave serves the Sultan. The girls did not so much seem accessory wheels to the general machinery as mere cogs to the wheels."[15] Similarly, *Moby-Dick* portrays the whaling ship itself, the *Pequot,* as a productive machine; its story details the technology of whaling and chronicles the corruption caused by Ahab's mechanistic obsession. Figures of machinery appear throughout the book; when Ishmael sees the skeleton of a whale he likens it to the intricacy of a textile factory. When one has harpooned a whale, he observes, the line connecting the harpoon flies out of the boat, and "to be seated then in the boat, is like being seated in the midst of the manifold whizzings of a steam-engine in full play, when every flying beam, and shaft, and wheel, is grazing you."[16] Melville grounded his novels in the machinery of the world, elegant but ominous.

Mechanical World in a Man-of-War

Most relevant to Melville's view of the *Monitor,* however, is his work about life in the navy, *White Jacket,* which taps his own experience. He joined the frigate *United States* (sister ship to the *Constitution*) in Honolulu in 1843 and sailed on it for more than a year, traveling to the Marquesas, Tahiti, Peru, Mexico, and Brazil before accompanying the vessel back to Massachusetts in 1844. This ship itself connects to the *Monitor* story: years later, the *United States* was captured at Norfolk the same night as the *Merrimack* and became the first ship of the Virginia Navy. In fact, one of the midshipmen on Melville's cruise was William Jeffers, later the notably utilitarian commander of the *Monitor.*[17] Aboard the *United States,* Melville found a rich library (stored in a cask) and easy access to newspapers. In addition to literature, he also found pettiness and brutality. The experience, combined with wide reading and borrowing from contemporary sources, went into his major work on naval seafaring, *White Jacket, or The World in a Man-of-War,* written in less than two months and published in 1850. Much of the novel involves not grand narrative but the daily routine of life aboard a sailing warship. *White Jacket* met with critical approval, and unlike *Battle Pieces,* it sold well.

White Jacket depicts both the people and the machinery of naval warfare in the pre-ironclad era. In this remarkable, almost ethnographic account, Melville chronicles life in the antebellum American navy; the book itself ranks as a valuable historical document of the late age of sail. "White Jacket" refers to the nickname of the narrator and his trademark garment. The subtitle, "The World in a Man-of-War," offers the metaphor of the ship as a microcosm, familiar and worn terrain today but still novel to the audience of 1850. *White Jacket* became best known for its strident, graphic portrayal of the barbarous ritual of flogging and its religious denunciation of the practice. It was later influential in Congress's decision to ban the punishment in 1851. Melville uses a mechanical figure for the ship's "general social condition": "The whole body of this discipline is emphatically a system of cruel cogs and wheels, systematically grinding up in one common hopper all that might minister to the moral well being of the crew."[18]

One incident in *White Jacket* prophetically links the book directly to John Ericsson and his inventions. While in harbor, the crew of the *Neversink* learned of the "lamentable casualty that befell certain high officers of state . . . all engaged in experimenting on a new-fangled engine of war" (130). This "engine" was Ericsson's *Princeton,* the tragedy the explosion of the Stockton gun that killed the secretaries of state and of the navy, among others. Indeed, the incident occurred in 1844 when Melville himself was aboard the *United States.*[19] As in the book, the ship fired a salute to honor the dead.

Violence of purpose pervades the man-of-war's every detail. It especially degrades the officers: "How were these officers to gain glory? How but by a distinguished slaughtering of their fellow-men" (208). Officers on the *Neversink* range from noble to useless to brutal (it is curious to think that Melville himself may have disliked the same, though younger, William Jeffers who so annoyed Keeler). But Melville's account emphasizes the lives of the ordinary seamen, describing their living, sleeping, and working conditions. Sailors aboard the *Neversink* keep journals (three other accounts of Melville's *United States* cruise survive), read books, write poetry, and recite memorized verse in their off-hours. They also build and maintain a history, retelling old stories of naval heroes like Stephen Decatur and "Old Ironsides . . . as Catholics do the wood of the true cross" (9).

As on the *Monitor* and the *Pequot,* these men formed not only an intellectual and social unit, but a productive unit as well. Again, Melville uses industrial analogies. White Jacket likens the ship to "something like life in a large manufactory" and finds himself "a long time rapt in calculations," simply trying to remember the various numbers that governed his life: the number of his mess, the ship's number, which hammock, which gun (35). The regimentation of the *Neversink* alludes to the growing "mechanism" in American life that later informed Melville's reactions to the *Monitor.*

Despite his aversion to the ship's constricting social structure, Melville admires the skills and crafts that support it. "Frequently, at one and the same time, you see every trade in operation on the gun-deck—coopering, carpentering, tailoring, tinkering, blacksmithing, rope-making, preaching, gambling, and fortune-telling." Although the man-of-war replicates the social structure and conflict of the world ashore, it is also a professional microcosm, a snapshot of the numerous activities that maintain a social body, floating or otherwise:

> From a frigate's crew might be culled out men of all callings and vocations, from a backslidden parson to a broken-down comedian . . . Bankrupt brokers, bootblacks, blacklegs, and blacksmiths here assemble together; and castaway tinkers, watch-makers, quill-drivers, cobblers, doctors, farmers, and lawyers compare past experiences and talk of old times. Wrecked on a desert shore, a man-of-war's crew could quickly found an Alexandria by themselves, and fill it with all the things which go to make up a capital. (74)

These trades, of course, had their own hierarchy. For Melville, however, the ranking had less to do with degrees of skill, education, or tradition than with physical positions and tasks. One chapter, "The Good or Bad Temper of Men-of-War's Men, in a Great Degree, Attributable to Their Particular Stations and Duties aboard Ship," maps individual character onto the ship's structure. The men of the tops work and relax in the rigging high above the deck; they enjoy space, breadth of vision, and intellectual pursuits, "lifted high above the petty tumults, carping cares, and paltrinesses of the decks below." White Jacket asks, "Who were more liberal-hearted, lofty-minded, gayer, more jocund, elastic, adventurous, given to fun and frolic, than the top-men of the fore, main and mizzen masts?" He believes only experience in the maintop could prepare him to write this "free, broad, off-hand and, more than all, impartial" account of man-of-war life, "meeting out to all—commodore and messenger-boy alike—their precise descriptions and deserts" (46–47).[20]

Life belowdecks, like life on the *Monitor,* was darker, quieter, and notably more industrial. *White Jacket* portrays the bowels as places of threat, mystery, and men of questionable motives. Architectural analogies drive the point home. The *Neversink* "is like the lodging houses in Paris, turned upside down; the first floor, or deck, being rented by a lord; the second, by a select club of gentlemen; the third, by crowds of artisans; and the fourth, by a whole rabble of common people." Some of the discomfort stems from depth: "a man-of-war resembles a three-story house in a suspicious part of town, with a basement of infinite depth, and ugly-looking fellows gazing out of the windows" (75). Darkness contributes fear as well. White Jacket likens the gunner to the archetypal worker in an artificial environment, "like a Cornwall miner in a cave, [he] is burrowing down in the magazine under the Ward-room, which is lighted by battle-lanterns" (67). In the depths of the ship, machinery and technical apparatus replace the proximity to heaven that White Jacket so loves about the tops. "By far the most curious department of these mysterious store-rooms is the armory, where the spikes, cutlasses, pistols, and belts, forming the arms of the boarders in time of action, are hung against the walls and suspended in thick rows from the beams overhead" (123–24).

One episode illustrates how these mechanical spaces degrade their inhabitants. One of White Jacket's sailor colleagues was promoted from his perch in the maintop down to the gun deck. His old mates from the tops went to visit him in the new surroundings, but "instead of greeting us with his usual heartiness and cracking his pleasant jokes, to our amazement, he did little else but scowl . . . he seized a long black rammer from overhead, and drove us on deck; threatening to report us if we ever dared to be familiar with him again." White Jacket attributes the change to immersion in the mechanical world of the gun deck. "It was solely brought about by his consorting with those villainous, irritable, ill-tempered cannon; more especially from his being subject to the orders of those deformed blunderbusses, Priming and Cylinder" (45). In general, he concludes: "A forced, interior quietude, in the midst of great outward commotion, breeds moody people. Who so moody as rail-road-brakemen, steam-boat engineers, helmsman, and tenders of power-looms in cotton factories? For all these must hold their peace while employed, and let the machinery do the chatting; they can not even edge in a single syllable" (46). Machinery was less likely than rigging and sails to make poets and philosophers out of sailors.

White Jacket informs our understanding of Melville's response to the *Monitor* by making it clear that the depredations of naval life existed before steam propulsion and iron hulls. Melville recognized that nostalgia for "navies old and oaken" simplified the past as much as blind praise for new technology simplified the future. Sailing ships were repositories of skill and dignity, but their own low machinations could also debase the serenity and high-minded leisure of the tops. *White Jacket* suggests that a steam-powered machine of war, especially one like the *Monitor* with its engines and living spaces underwater, must be a churlish place indeed.

"Among the Trades and Artisans"

How would the world in a man-of-war mix with the appetites of steam-powered machinery? Who would dominate the outcome—brutal professionals, high-minded sailors, or moody machine tenders? These were the questions the *Monitor* raised for Melville, and he addressed them in "A Utilitarian View of the Monitor's Fight," the richest of the ironclad poems and among the best of *Battle Pieces.* Irony pervades the poem: poetry, the vehicle of beauty and love, here mimics warlike mechanism. The poem itself works like a machine: "Plain be the phrase, yet apt the verse / More ponderous than nimble." Familiar rhyme, like familiar navies, no longer obtains; now cadence and diction march with a mechanical dissonance:

> For since grimed War here laid aside
> His Orient pomp, 'twould befit
>   Overmuch to ply
> The rhyme's barbaric cymbal.

The poem sees fighting machines like the *Monitor* as ending war's pomp, reducing battle to a technical exercise:

> Hail to victory without the gaud
>   Of glory; zeal that needs no fans
> Of banners.

New methods of warfare will belong to new practitioners:

> . . . Plain mechanic power
> Plied Cogently in War now placed—
>   Where War Belongs—
> Among the trades and artisans.

War clanks mechanically along, but for human participants battle becomes more dangerous, "deadlier, closer," if less grand, but now under the "calm, 'mid storm" of engineering calculation rather than base emotions.

> No passion; all went on by crank
>   Pivot, and screw,
> And calculations of caloric.

The poem articulates the significance of Hampton Roads: "The clangor of that blacksmiths' fray . . . Resounds this message from the fates."

> War shall yet be, and to the end;
>   But war-paint shows the streaks of weather;
> War shall yet be, but warriors
> Are now but operatives; War's made

> Less grand than Peace
> And a singe runs through lace and feather.[21]

This message resonates with the experiences aboard the *Monitor*. Its crew recognized the new role of the "trades and artisans" in naval warfare. As Ericsson wrote to engineer Stimers, "This system [the *Monitor*], we cannot omit to observe, has called into action a new element of success in naval conflict viz. engineering skill."[22] Melville's emphasis on "war-paint," and "lace and feather," captures the sense that the social experience of war is as much about symbolism and appearance as it is about fighting. Indeed, Keeler's own life aboard the ship, and the lives of much of the crew, certainly seemed "less grand than peace," their own anxieties reflecting the poem's image of warriors as "operatives" in factories. Recall Keeler's early impression, that "there isn't danger enough to give us glory," and his awareness that fighting behind thick armor did not necessarily imply bravery.

On first glance this poem, with its celebration of "calculations of caloric" (an allusion to Ericsson's caloric engines), seems one that even John Ericsson might have hung on his office wall. Rhyming "heroic" with "caloric," it implies that Ericsson the calculator was as much a warrior as Worden, Greene, Keeler, or the *Monitor* crew. Indeed, William Conant Church precisely echoed this notion in his biography of Ericsson: "Improvement in warlike appliances, and the professional study of war tend to destroy the demoralizing sentiment of personal hostility toward a public enemy, encountered in battle." For Church, as for Ericsson, mechanical warfare would make war more civil by enlisting only calculation, not emotion and experience. "There is nothing to produce this sentiment [hostility] in the breast of a man absorbed in scientific manipulation of warlike machinery, or in solving problems of logistics, strategy, and tactics at such a distance from an enemy that the loss of life, to which he contributes, figures in personal consciousness only as one item in the statistics of a great contest, or among the sounding phrases of a war bulletin."[23]

On a closer reading, however, "A Utilitarian View" subverts the triumph of military engineering. Ericsson and Church sincerely believed the gospel of mechanical war, but Melville's voice of the "utilitarian" equivocated. In fact, the word "utilitarian" had negative connotations for Melville, as it did for Dickens, whom Melville admired. Elsewhere in *Battle Pieces* he uses the word pejoratively. Perhaps, as one scholar has suggested, Melville is lamenting his own obsolescence, not that of wooden warships.[24] Recalling *White Jacket*, with its admiration of the many skills that go into the workings of a sailing warship, Melville questions the notion that war is now placed "among the trades and artisans." In *White Jacket* war already engaged varied skills—indeed, skilled workers gave the warship its virtue. The mechanical spaces, with their dark, submarine depths, were spaces of fear and corruption, not cool calculation. There, proximity to machinery produced dangerous, antisocial men, not thoughtful, detached engineers. Moreover, the "war-paint" and "lace and feather" that Melville portrayed on the *Neversink* nauseated him with their unthinking discipline and cruelty in the guise of order, an abhorrence unlikely to be undone even by the shocking *Monitor*. Nor, in Melville's view, is traditional naval heroism about to die, anyway: "In the Turret," as well as the experiences of the *Monitor* crew after Hampton Roads, make it clear that heroism has a place even in the age of the welded war machine. No, "A Utilitarian View of the *Monitor's* Fight" proclaims no simple admiration for a war conducted by technical experts. Melville's strongest statement in the poem is duality, his recognition that the machine will cut both ways. The poem shares an essential element with Hawthorne's article and Keeler's letters, an element John Ericsson totally lacked: irony.

Iron and Irony

Melville's poem highlights the contradictions surrounding the *Monitor's* early stirring of the mechanization of war. That the poem is the strongest of his set on the ironclads suggests the appropriateness of the ironic voice. Ironically, engineering calculation produced inhuman conditions. Ironically, while going to meet the enemy, the impregnable ship almost killed its crew. Ironically, those concerned about the status of their heroism ended up becoming heroes. Ironically, the powerful war machine was neutralized by its value as a public symbol. Ironically, the heavy armor did as much to imprison as to protect its inhabitants. Keeler's own sense of irony increased as the discrepancy grew between the *Monitor's* military effects and its display value. Hawthorne amplified Keeler's perceptions in an account structured by irony: the contrast between what looked like a "gigantic rat trap" on the outside and the comfortable quarters inside. Indeed, Hawthorne found sarcasm the only appropriate mode to convey the contradictions of the war—especially those of the mechanical weapons that he saw from his strange, detached vantage point.

The *Monitor* introduced the world to the specter of industrial warfare. Participants and observers wondered how it would affect the experience of combat. Fighting behind impregnable armor could strip victory of its glory. Utilitarian engineers might usurp the splendor of battle from the officers' experience. A new moral economy of war would reallocate expertise and credit. The *Monitor's* crew felt these changes during their own immersion in the machinery. Ericsson thought he understood the transformation, but his very insistence suggested that uncertainty underlay his calculations. As Keeler, Hawthorne, and Melville articulated, war machines *can* reduce human participation, but they also dramatically alter the nature of the experience for those who do participate. Unlike the imaginary mechanical war that, in Hawthorne's words, leaves a field "strewn with broken engines," Keeler experienced a war that left a field (and a sea floor) strewn with broken bodies. Despite the visibility of engines and armor, fields and bodies still dominated the American Civil War. Those same images, and not "calculations of caloric" would dominate the conflict that was to begin in 1914.

Conclusion

Mechanical Faces
of Battle

 That, then, was the war he had been so eager to get into at one time, the opportunity for basic, hot-tempered, hard-muscled heroism he regretted having missed. There had been plenty of death, plenty of pain, all right, and plenty of tooth-grinding stoicism and nerve. But men had been called upon chiefly to endure by the side of the machines, the terrible engines that fought with their own kind for the right to gorge themselves on men. Horatio on the bridge had become a radio-guided rocket with an atomic warhead and a proximity fuse. Roland and Oliver had been a pair of jet-driven computers hurtling toward each other far faster than the flight of a man's scream. The great tradition of the American rifleman survived only symbolically, in volleys fired into the skies over the dead in thousands of military cemeteries.

Kurt Vonnegut, *Player Piano*, 1952

At the close of the Civil War in 1865, returning soldiers and civilians surveyed the damage, mourned the dead, and tried to put the horror behind them. Americans were "willing and even anxious to thrust into shadow all things martial."[1] The navy, hardly the beneficiary of a revolution, entered its "dark ages" from 1869 to 1883, as the military radically demobilized from war. Gustavus Fox and Gideon Welles dismantled the formidable wartime force they had built, reducing its numbers by more than 80 percent. Forgetting many of the war's lessons and accomplishments, the navy returned to its dated strategy of river and harbor defense and to ships of canvas and wood. Some ships retained steam for auxiliary power, but their commanders faced penalties for even using the machinery and incurring the extra costs of coal. In 1869 Benjamin Franklin Isherwood lost his job as engineer-in-chief, humbling the engineering corps he had tirelessly promoted. "So complete was the restoration [of the old navy]," writes historian Robert Albion, "that a visitor returning in 1870 after ten years absence might never have guessed that the navy had passed through any war at all, much less a war that had hastened a revolution in naval architecture and thoroughly demonstrated the inadequacy of the strategic theories in vogue on the eve of the conflict."[2] The *Monitor,* too, largely faded from memory. Popular enthusiasm for ironclads waned as much of the monitor fleet rusted in storage in Philadelphia. Few of the pamphlets and speeches commemorating the *Monitor* date from the ten years after 1865. In 1874 *Monitor* captain John Worden petitioned Congress for prize money for his crew, but the request went nowhere in the face of angry Southern opposition.[3]

By the 1880s the "mechanism" that Hawthorne and Melville had observed matured into a newly powerful industrial culture. In that decade, partially as a nostalgic response to the competitive, materialistic age, popular interest in the war began to revive. The publication of Grant's memoirs stirred popular interest in 1885, and Admiral David Dixon Porter's history of the navy in the Civil War appeared the following year. Veterans, many dissatisfied with their

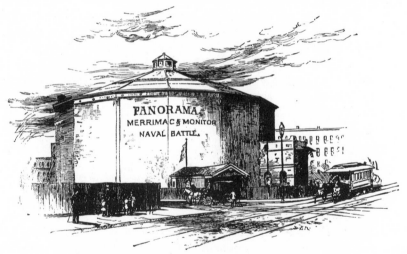

CORNER MADISON AVENUE AND FIFTY-NINTH STREET, NEW YORK.

The *Merrimac* and *Monitor* Panorama, a tourist attraction set up in New York City in the 1880s. "I visited New York this summer to see one or two of my old commirads," wrote John White, a member of the *Monitor* crew, "and went to see the panorama of the fight. The location of the vessels is not at all correct." White was disgusted with the tour guide, who made Lieutenant Samuel Dana Greene out to be a hero. Courtesy of the Gardner Weld Allen Collection, Widener Library, Harvard University.

own peacetime progress, flocked to associations and celebrated memorials, lionizing what seemed to be a morally simpler time.[4] Further prize bills for *Monitor* crew were introduced in Congress several times from 1882 to 1884 (they were still never able to overcome the opposition, which came both from the South and from those who believed the *Monitor* crew had done nothing unusually heroic, although one bill did pass in the Senate).[5] Nonetheless, a wave of publications, speeches, and memoirs swept the country. In New York City a "panorama," a circular diorama the size of a city block, recreated the conflict between the *Monitor* and the *Virginia* for the public. The program for the multimedia tourist attraction declared, "The observer who stands in the center of it cannot discern where the real joins the representation, nor can he fail to imagine himself on the very spot, the conflict going on about him."[6] As always with the *Monitor,* symbolism and history were not easily disentangled.

The Historical *Monitor*

In addition to popular attractions, a singular publishing event boosted the Civil War revival of the 1880s. In 1887 *Century* magazine collected an array of firsthand accounts of the war, publishing them as a series of articles and then as a four-volume set, *Battles and Leaders of the Civil War. Century*'s circulation doubled. The series, which included accounts by Generals McClellan, Sherman, Grant, Johnston, Longstreet, and many others, was reprinted in 1956 in anticipation of the Civil War centennial anniversary.[7] Nine articles concerned the construction of the ironclads and their fight at Hampton Roads. Eyewitness accounts from both sides retold the battle, engineers John Brooke and John Porter recalled the conversion of the *Merrimack,* Ericsson described the construction of the *Monitor,* and the magazine reprinted the text of the *Monitor* partners' negotiations with the Ironclad Board. The structure of the *Century* articles established the canonical narrative; some historians still rely on them as authoritative. Not one of the *Century* articles described the long summer up the James River.

We must understand the *Century* articles in the context of their time. Participants saw in them an opportunity to settle the questions that had remained open for twenty years. For example, Samuel Dana Greene, now 44 years old, wrote his own article for *Century,* "In the *Monitor* Turret," which narrated Greene's experience as the *Monitor*'s youthful executive officer and as commander of the vessel after Captain Worden's injury.[8] He described the difficulty of fighting in the turret, the inadequate powder loads for the guns, and the circumstances that ended the battle.

After sending his account off to the publisher, Greene promptly shot himself in the head.

Unlucky Greene was reacting to a debate that had continued privately since the war and that increasingly focused on his own behavior. Since the March day in 1862, questions had remained, and uncertainties continued to plague the *Monitor* veterans. Captain Worden had never written an official report of the battle, nor any account, until 1868, when he finally penned a letter to Secretary Welles: "Recently learning that Lieut. Cmdr. Samuel Dana Greene has been annoyed by ungenerous allusions to the fact that no official record existed at the Department in relation to my opinion of his conduct on that occasion, I desire now to remedy a wrong, which I regret should so long have existed, and to do justice to that gallant and excellent officer."[9] For the failure to destroy the *Virginia,* Worden blamed insufficient powder charges, lack of training for the crew, and the novelty of the vessel, charges that Greene echoed in his later account.

Open Questions, Closed History

After the war, John Ericsson and his supporters believed the questions surrounding the *Monitor* were closed, but the controversy had never quite died. In 1874 Gustavus Fox began a private attempt to fully document the Battle of Hampton Roads and inadvertently struck a nerve. He asked Catesby ap R. Jones, the lieutenant in charge of the *Virginia* during the fight, to write his own account of the encounter. Jones, in an article later published, accused the *Monitor* of running away in defeat—a claim consistently made by Southern participants.[10] Fox forwarded a copy to Ericsson, who angrily replied, "I am quite at a loss to understand why you have opened a fresh discussion about the *Monitor* and the *Merrimack* fight, so happily disposed by several patriotic writers to the satisfaction of the country—I may say to the satisfaction of the whole world." In handwriting shaking with his fury, Ericsson accused Greene of losing the battle. For the first time on paper, he admitted that the fight with the *Virginia* ended in something less than a total victory for the *Monitor.* "No one knows better than yourself the shortcomings of that fight, ended at the moment the crew had become well trained and the machinery got in good working order. Why? Because you had a miserable executive officer who in place of jumping into the pilot house when Worden was blinded ran away with his impregnable vessel." Were Jones's account to be published, Ericsson wrote, "its effects will be more damaging than probably any incident of my life."[11]

Ericsson had reason to fear reopening the matter. Despite the fragile closure, his own heroic position was less than ironclad. Now, ten years after the war, Fox was hearing from several men associated with the *Monitor.* Gideon Welles, also writing to Fox, declared, "The failure of the contractors to deliver the *Monitor* for more than a month" after the agreed due date left Hampton

Roads vulnerable to the *Virginia* and caused disaster for the *Congress* and the *Cumberland*.[12] Before his death, Samuel Dana Greene defended his command of the *Monitor* to Fox, writing that he did not pursue the enemy ironclad because the wounded Worden ordered him to hold back and defend the *Minnesota*. Greene argued that the *Monitor's* technical deficiencies determined the outcome of the fight: the *Virginia* ran away faster than the *Monitor* could pursue, and the pilothouse prevented firing ahead while giving chase. Alban Stimers, Ericsson's man aboard the *Monitor,* concurred that he had advised Greene to hold the *Monitor* back.[13] In the face of these open questions, a dichotomy emerged between personal heroism and technical success: if the *Monitor* was the perfect war machine, its failure to achieve victory could only be caused by imperfect men. If the vessel was technically deficient, however, Greene and the crew had heroically overcome its problems to face the enemy.

These were not arcane questions. Greene's suicide exposed the depths of emotion that still surrounded the *Monitor's* performance. His death, and the appearance of the *Century* articles, resonated with the survivors. Many wrote to Frank H. Pierce, a New York collector of *Monitor* memorabilia. Most saw the suicide as directly related to questions about Greene's heroism. Some blamed Greene's youth and indecision for breaking off the engagement after Worden was wounded. Frank Butts, who was not aboard during Hampton Roads but who joined the *Monitor* in October 1862, questioned Greene's behavior. "I believe his own mortification and failure to cover this point was the cause of his suicide." William Durst wrote that Greene was considered a coward by the crew. John M. White, a fireman, reported that after the encounter "the crew stood around in groops on deck and omong themselves accused Greene of being a coward." White eagerly visited the battle panorama in New York but turned away in angry disgust when the tour guide portrayed Greene as a hero. Louis Stodder merely considered Greene's account overly dramatic. Samuel Taylor, who became Ericsson's secretary and assistant in 1864, went so far as to suggest that the crew of the *Monitor* considered the battle an unmitigated defeat until they heard cheers from shore and Fox came aboard to commend them. For this reason, Taylor reported, a portrait of Gustavus Fox hung on the wall of Ericsson's home, its only ornament.[14]

In general, enlisted men attacked Greene, whereas officers defended him. The infirm Keeler, now settled in Florida, also corresponded with Pierce (when Keeler was too ill to write, Anna wrote for him). After the *Monitor's* sinking, Keeler and Greene (and several other *Monitor* officers) had served together on the blockade steamer *Florida* and had become friends. "I knew him well," wrote Keeler, "and I know full well that there was nothing like cowardice about him. Who or what has started these rumors I don't know. I only know them to be false. I cannot help but think that this rumor had reached him &

was one of the causes of his untimely death." Keeler thought the crew's attacks on Greene stemmed from distorted memories, whereas his own version had a documentary base. "I kept all my experiences written up in the form of letters to my wife. These were all preserved and I have them bound in book form to refer to."[15] Keeler died February 27, 1886, just two weeks after his last letter to Pierce.

In *Century*, the issue of living conditions surfaced once again. Greene echoed the sentiments of his comrades that "probably no ship was ever devised which was so uncomfortable for the crew." For Ericsson, as before, without mechanical expertise one was not fit to judge even the unpleasantness of the experience. "If this opinion were well founded," he countered, "it would prove that submerged vessels like the *Monitor* were unfit to live in" (which of course they were). Ericsson supported his statement with data, quoting a report by the chief of the Bureau of Medicine and Surgery in 1864, which showed that the rates of sickness aboard the monitors were lower than they were on wooden ships.[16] Similarly, Greene blamed the *Monitor*'s poor gunnery on the difficulty of operating the engine that rotated the turret (its momentum caused it to overshoot its mark), but Ericsson claimed that the crew had not properly maintained the engine on the trip from New York, this despite the "waterfalls" surging down on it from the leaky hull.

Controversy centered on the sinking as well as the battle. In the *Army and Navy Journal*, Ericsson charged the crew with being drunk during the fatal gale and unable to save the ship. Here and in *Century*, he publicly blamed Greene for the tragedy. "The important question must remain unanswered," he wrote in the latter, "whether in the hands of an older and more experienced executive officer the *Monitor*, like the other vessels of her type, might not have reached Charleston in safety." Louis Stodder vigorously protested this characterization of events, writing to Pierce that Ericsson "covers up defects by blaming those that are now dead." Instead, Stodder focused on technical problems, attributing the vessel's loss to the overhang between the upper and lower hulls: because of the shock the overhang received in a heavy sea, it parted from the lower hull, dooming the vessel. Both Geer and Keeler had reported the terrible slam of the overhang prior to sinking. Ericsson, in contrast, claimed that the "waterfall" around the turret came through a space that should not have been there: he designed the turret to rest on a huge brass ring, its flat metal face sealing under the weight of the turret to keep the water out. The *Monitor*'s officers, he asserted, had put oakum (hemp rope impregnated with pine tar) between the faces in order to improve the seal, which in actuality destroyed it: "The accident was simply the result of putting oakum under the turret."[17]

Historians have generally credited Ericsson's absentee version of the sinking, blaming the crew's installation of oakum. Evidence equivocates on this point, however. Witnesses recounted water rushing in from the hawse hole in the anchor well and commented on the threat, if not the actuality, of the two hulls parting. Frank Butts reported that oakum had been installed in the turret, but Louis Stodder wrote that he, "as navigating officer, should see that the turret was properly prepared for sea *and it was* and no oakum under it." The seal did leak, but "the amount of water that came in around the turret was hardly perceivable and cannot be attributed to that cause."[18] At other times crew members said that the turret seal was faulty and needed additional packing beyond the metal-to-metal contact. On the *Monitor*'s first trip, engineer Alban Stimers actually *complained* that "this vessel came to sea with out a particle of oakum. The bottom of the turret was not backed and of course it leaked by the bucketfull, the water pouring right down on the berth deck. The hatch also leaked so that everything was wet and uncomfortable." John Rodgers, whom Ericsson praised for weathering the gale in the *Weehawken,* did seal his turret base with oakum on his first trip. "Our turrets are never quite tight," Ericsson reluctantly conceded. "But the leak (expected by the constructor) is insignificant and remains *permanent* [and] hence can never be dangerous," he concluded, in unusually imperfect logic.[19] George Geer reported that packing had been installed after Hampton Roads but had rotted away over the summer.[20] Geer's observation raises the question whether it was replaced or removed during the October overhaul.

Whatever the actuality, the rhetorical configuration remains the same: debates over both the success of the *Monitor* in battle and its failure at sea counterpoised the heroism and experience of the machine's operators with the technical competence of the constructor and the engineer. Thus occurred the very debate Melville had foreseen, one common to accident investigations today: human error versus mechanical failure.

The popular *Century* articles helped establish the legend of John Ericsson, the lone inventor pitted against conservative professionals. The *Army and Navy Journal* continued to support Ericsson and his monitors, and the inventor continued to make his history. In 1876, having been "inadvertently" omitted from the Centennial Exposition (because of a disagreement with the organizers of the Naval Exhibit), Ericsson published his own pamphlet describing his inventions, from the *Princeton* through the monitors and his new solar engines.[21] Soon after the *Century* articles appeared, he invited William Conant Church to write the admiring biography, although only after destroying his personal diaries.

Ericsson died on March 8, 1889, one day before the twenty-seventh anniversary of the *Monitor*'s fight at Hampton Roads. His funeral brought his name, by then largely forgotten, back to public attention. A modest procession bore his casket to a vault in Manhattan. The Department of the Navy, always unsure of how to deal with the inventor in life, decided to send the body to

Sweden for burial, responding to a request from the Swedish government. For the navy, elevating Ericsson to heroic status neatly accorded with an image of revived technological innovation. The "new navy," its dark ages over, had in 1883 begun to be the instrument of the country's budding global ambitions. On August 23, 1890, "John Ericsson Day" in New York, the body was ceremoniously loaded on a navy warship for the trip across the Atlantic. Thousands turned out, including the secretary of the navy, William Conant Church, Cornelius Bushnell, Samuel Taylor, Captain Worden, six other *Monitor* men (Stodder, Hannon, Anderson, Geer, Nichols, and Toffey), and twelve actual monitors, including the *Passaic*. The *New York Times* reported "an awakened appreciation of the qualities of a man so retiring and unobtrusive in his personality that to the mass of citizens who turned out to do him homage he was an entire stranger."[22] Received with comparable pomp in Stockholm, the body was finally interred near Ericsson's birthplace in Filipstad. Just a few months later the Church biography appeared, expanding a series of written sketches Church had published in *Scribners* magazine. At last, John Ericsson received the recognition he craved. His funeral, combined with the *Century* articles and Church's book, created the standard form of the *Monitor*'s history: the genius inventor, the navy's conservatism, the frenzied building, the revolutionary battle, and the heroic loss. The legend of the *Monitor* emerged from a blend of contemporary reporting, public debate, and historical commemoration.

The *Monitor* Wreck: Artifact and Icon

For more than a century, the legend sustained itself on its own, absent the ship itself. That changed in 1973, when the *Monitor* returned to inform the legend. A group of scientists discovered the wreck sixteen miles off the Cape Hatteras lighthouse in North Carolina. The site was declared a National Marine Sanctuary (a distinction usually reserved for coral reefs and kelp beds), listed in the National Register of Historic Places, and designated a National Historic Landmark—unusual honors for a place, or rather a thing, more than two hundred feet below the surface of the ocean.

In the ensuing decades, archaeologists have made numerous visits to the wreck, surveying its condition and recovering artifacts. They found that the ship capsized as it sank and that the turret came loose and landed first on the bottom. The hull then settled upside down, resting on the turret, which props it up off the seafloor. Ironically, the most lasting design feature of the *Monitor,* the turret, is now responsible for its rapid decay, because it allows water to flow freely underneath the wreck. Archaeologists have recovered the anchor, the propeller, a signal light, and even an intact jar of relish, along with a number of small objects, and are restoring them for museum display.[23] Citing deterioration of "crisis" proportions, the National Oceanic and Atmospheric Administration (NOAA), which oversees the wreck site, has recently proposed a project to recover significant parts of the vessel, including the massive turret and the Dahlgren guns.[24] The proposal has occasioned yet another in the *Monitor*'s many rounds of press coverage, including a full-page article in *Newsweek* magazine and a picture on the cover of the *New York Times* "Science Times."[25] The *Monitor* makes as compelling an icon in the late twentieth century as it did in its own time.

Discovery and exploration of the wreck have been added to the standard accounts, a final chapter following the sinking. But the site and the artifacts have added little to a strictly historical understanding of the *Monitor*. Indeed, the published literature on the wreck focuses less on the ship itself than on the processes and technologies of search, diving, and excavation. One critic even called the results of the *Monitor* explorations "little more than a handmaiden to history . . . historical trivia."[26] This accusation is surely unwarranted, for archaeologists have concentrated on assessing the wreck first as a thing, rather than as a historic icon. Understanding the wreck's state of decay, its possibilities for recovery, and its potential for conservation is a necessary first step toward studies that would allow the material *Monitor* to contribute to the historical *Monitor*.

What can the *Monitor* as a thing contribute to the *Monitor* as history? A more accurate picture of the *Monitor* as built might clarify Ericsson's design technique. Technical enthusiasts would like to understand the precise plan of the vessel, because Ericsson never made a complete set of drawings, and numerous modifications were undoubtedly made during construction and operations.[27] Further exploration of the wreck may also help pinpoint the precise cause of the sinking: did water rush in through the joint between the hulls or through the leaky turret base? In other words, what was at fault for the loss: design, maintenance, or ship handling? Ericsson, navy engineers, or line officers? More important for the purposes of a study like this one, however, excavation of the ship's interior might reveal details about life aboard the *Monitor* to complement the written sources. How did the crew adapt to the conditions aboard the ship? Did they modify it to make it more habitable? Keeler's stateroom, his writing desk, his safe, and perhaps even Anna's letters are down there somewhere.

Although it might contribute to the historical *Monitor*, the wreck should also be considered on its own. Whereas an image of the *Monitor* works through metaphor, evoking the ship by its likeness, an artifact's connection is metonymic, evoking history by physical proximity. Artifacts from the *Monitor* were literally adjacent to the historical events. This has been found to spark interest of an intensity comparable to the original enthusiasm for the ship (as a comparison, consider the ongoing cult of the *Titanic*). The *Monitor* wreck establishes a material connection to the ship's human actors, including designers, shipyard workers, engineers, enlisted men, and officers. Perhaps even the *Monitor*'s visitors left physical traces. Whether recovering such traces is worth the expenditure of public or private funds, however, remains a matter of debate.[28] Even today, then, the *Monitor* raises questions about the relationship between appearance, conveyed through images, and experience, in contact with things.

Technology ties the historical *Monitor* with the archaeological one; stories and images of the ship now appear alongside those of the submarines, the robots, and the skilled divers exploring it in the depths. The machine that once stood simultaneously for progress and destruction now stands both for preservation and for decay. What once spoke success also testifies to failure—the ship that fought but did not float.

Marrying Machines and Guns

The eerie, submarine qualities of the wreck and the strange divers and submarines required to access it remind us of another aspect of the *Monitor*'s enduring significance: the human experience of technology, particularly in war. The local, machine-mediated experience aboard the *Monitor* reflected larger trends. Soldiers' expectations in the Civil War evolved from a series of heroic contests to an incessant, machine-like battle. The enthusiasm of 1861 and 1862, when familiar notions of courage and heroism still defined the soldier's world, contrasts with the crushing, grinding war of 1864 and 1865, a world of trenches, stasis, and high casualty rates. "Forces at play in the Civil War," writes Gerald Linderman,

> signaled something far more potent than combat's cost in lives: that the very *nature* of combat did not fit, and could not be made to fit, within the framework of soldier expectations. Forces of change and novelty made themselves felt less dramatically and drastically, but they slowly chipped away at soldiers' resolve, and their results were over time more profoundly dispiriting. Ultimately, they led many to the realization that they could not fight the war they had set out to fight.[29]

Linderman describes the world of the soldier, not the sailor, and chronologically the ironclad belongs to the early, glorious war rather than the later, grinding one. Even so, his interpretation of soldiers' experience helps explain the lingering impact of Hampton Roads: it hinted at changes to come, not just for the war at hand but for mechanized conflict in general. The *Monitor*'s history, and the experience of its crew, compressed into a historical moment the tensions within mechanical warfare that helped define the industrial world.

Technological images did not dominate the Civil War as they did later conflicts, so the anxieties remained comparatively below the surface. For contemporary and historical observers, the words of Keeler, Hawthorne, and Melville went largely unnoticed. Daniel Aaron, in fact, has called the Civil War "the unwritten war" because no flood of literary expressions (as opposed to memoirs and histories) seized upon its horrors to question the relationship between industrialization and warfare. Indeed, no American epic emerged from the war at all; even *The Red Badge of Courage* and the film version of *Gone with the Wind* both appeared decades later and did not concern themselves with questions of industrialism and war. Aaron argues that writers' dispositions, literary technique, and the reading public were unable or unwilling to engage the war because they resisted discussing the central question of race.[30] We might extend this resistance to industry and technology: despite the reservations of Melville and Hawthorne, Americans in the nineteenth century did not question the nascent association of industrial technique with organized violence.

As Paul Fussel observed, it was the twentieth century that opened with the marriage of the word "machine" to the word "gun."[31] In fact, World War I revealed disturbing and unexpected continua with the American Civil War. Machinery, earthworks, and death continued to entangle as though unmindful of the fifty intervening years. Like the Civil War, World War I arrived on the heels of decades of celebratory progress, when the Western world came to believe that advances in science and technology promised the eventual perfection of human society. Unlike the earlier war, however, World War I occasioned reevaluations of that worldview. People began to realize that not only could death be the fruit of progress, but that mechanized destruction might always accompany industrial production, "becoming, like the telephone and the internal combustion engine, part of the accepted atmosphere of the modern experience."[32] A generation of writers, including Robert Graves, Wilfred Owen, Siegfried Sassoon, and Erich Maria Remarque, wrote of the experience of war in the industrialized world. They settled on irony as the mode that expressed the disparity between the experience of war and its public languages. "One of the cruxes of the war," writes Fussel, "is the collision between events and the language available—or thought appropriate—to describe them. To put it more accurately, the collision was between events and the public language used for over a century to celebrate the idea of progress."[33]

The armies of World War I deployed new technologies of artillery, poison gas, and machine guns to generate a field of carnage unparalleled in its modern efficiency. The place of the individual soldier within this large system led to the sense of powerlessness and longing that Graves, Owen, and others expressed in their writing. Mirroring the strangeness of the *Monitor,* participants in the trenches lived in a mythological world that imbued small details with significance, even superstition. "That such a myth-ridden world could take shape in the midst of a war representing a triumph of modern industrialism, materialism, and mechanism is an anomaly worth considering."[34] Industrialized war begs an ironic reading of historical experience and links the writers of World War I to other chroniclers of twentieth-century war (especially the highly ironic Joseph Heller, Thomas Pynchon, and Michael Herr), but it also evokes the irony with which Keeler, Hawthorne, and Melville viewed the *Monitor*.[35]

Consider, for example, the social and psychological dimensions of enclosure. As in the Civil War, the technology of World War I was not merely systemic, but also vehicular. Machines that enclosed men came to signify escape from the deadly systems, all the while carrying their own terrifying potential. Tanks, with their confining armor, their turrets, and (under favorable circumstances) their mobility, seemed like ironclads on land. Tank crews commented on their "iron monster" and the "absence of any visible human agency," an emptiness that generated an uncanny animism. "There was something inexorable and inhumanly purposeful about them."[36] Like the *Monitor*, tanks combined movement with enclosure, and also like the *Monitor*, they prompted debate about the effects of machinery on warfare. Military theorist J. F. C. Fuller called tank warfare "sea warfare on land." Proponents argued that tanks would restore lost agency, mobility, and speed to the battlefield, so that war would "again become more of an art and less of a business." Opponents saw tanks as dangerous, unreliable mechanical monstrosities that were beneath the dignity of professional soldiers.[37] The contrivance simultaneously reduced war to grinding factory work and recast it, recovering a space of prowess and courage within the machine.

Nonetheless, the tanks of World War I did not produce a new breed of technological heroes: that role fell to the airplane. As cultural icons, fighter pilots represented individuality, agency, and mobility in the midst of the brutal, collective war. Like the *Monitor* crew, air aces first became heroes in the eyes of frontline soldiers, who, "craning their necks to watch the battles above the trenches," witnessed their aerial battles and saw the pilot "imagined (benignly) as a mythic creature who restored perspective, consequence, and willfulness to the war as a whole."[38] Aces embodied the kind of war many on the ground had envisioned when they enlisted: a knightly one-on-one battle of skill and courage, ending in an honorable public death for the loser and glory for the victor. Life in the trenches—tedious, squalid, and collective—had none of the individual agency of the aerial joust. There machines rekindled the possibilities of heroic action, whatever the substance of their contribution to the war effort. More precisely, the social circumstances surrounding the technology bred a military elite. But even that elite reflected industrial ideals. The air aces represented not traditional military heroism but "sober, utilitarian middle-class values" more akin to bourgeois movie stars than aristocratic heroes (fewer than 10 percent of the German aces were nobles).[39] Heroism increasingly meant mastering the machine as much as destroying the enemy.

And mastery was essential, for the machines brought their own risks. New modes of fighting meant new modes of death, uniquely horrible. As William Keeler was only too aware, death-dealing technology has a central, ironic feature: the inherent danger of the machine for its operator. To "shoot down" another airplane often meant disabling the machine that would, in turn, kill its crew. Where the lives of the *Monitor* crew depended on the integrity of ventilators and blower belts, the life of the fighter pilot depended on the integrity of the airplane's structure and engine. Disabling either could kill the inhabitants and count as victory. Mechanical warfare defined an ironic relationship to the machine: simultaneously protector and weapon, potential savior and potential tomb. A grim statistic confirms the fact: more than half of all British pilot deaths occurred in training, with the enemy nowhere in sight.[40]

Even the lumbering airships and bombers shared imagery of enclosure with the *Monitor*. Zeppelins, which appeared above London in 1915, seemed all the more ominous because of their opacity, the crew hidden inside some cavernous, moving machine. "That the streamlined zeppelin itself seemed inaccessible, closed, without showing even a trace of the crew," writes Peter Fritzsche, "added to the sense of unknowable, possibly extraterrestrial power."[41] Mastering the cumbersome, unwieldy dirigibles presented a challenge to the crew comparable to that encountered in combat, "Rather than exerting dominion over nature, zeppelins fought a losing battle against not only enemy fire but the North Atlantic's inclement weather. The story of their wartime service is one of unequaled vulnerability and failure: nights were 'pitiless'; battles were 'horrible'; 'heaven and earth are alive to destroy us.'"[42] The German nation venerated the zeppelin's human operators, warriors who simultaneously battled the enemy, the elements, and the machine. Heroism meant "resolve in the *face* of technology, not the *power* of technology."[43] In these passages one might easily replace the word "zeppelin" with the word "monitor."

Most of the literature dealing with technology and the experience of war focuses on World War I on the ground and in the air. Naval historians have rarely addressed general questions of new technology and participants' experience. Yet naval technologies, especially the submarine, did raise issues of enclosure, agency, and expertise similar to those that surrounded the *Monitor*. The term "iron coffin," which was applied to the *Monitor*, became a routine epithet for submarines. One observer in 1902 noted the argument that "the effect of potted air on the crew would be disastrous." World War I submarines took the claustrophobia and bad air experienced aboard an ironclad to the extreme (adding diesel fumes and battery gasses to human effluents). Crews' endurance as much as that of the machines themselves set the limits on cruis-ing range.[44] During World War II scientists finally began to investigate the effects of heat, noise, motion, carbon dioxide, and unusual sleep cycles on submarine crews. "The unavoidable rigors of life on a submarine patrol," ran one analysis, "subject men to conditions of severe psychological and physio-logical stress and demonstrate an endurance that bears testimony to the remarkable capacity of the human organism for adaptation, for compensa-tion, and for acclimatization to conditions."[45] New disciplines of industrial culture replaced traditional skills and rituals of military life, making war a matter of specialized knowledge, artificial life support, and machine control.

Unmanned Weapons

As these brief examples show, the *Monitor* is not unique in the history of military technology. Technologies of war have always raised questions of power, failure, enclosure, protection, ennoblement. From gunpowder to cruise missiles, new weapons were derided as cowardly or unmanly ways of fight-ing, or praised as heroic mastery over nature and machines. Dreams of the impregnable defense, whether stone, iron, or electronic, are at least as old as armor; their technological extreme, the fully autonomous weapon, replaces humans altogether. The *Monitor* marks an event in the cultural history of those dreams, a point on the continuum from the armored knight to the unmanned fighter planes now in development. The question remains: whom do they unman?

Attempting to answer this question leads us to plot the *Monitor*'s career along new axes, to generate a history of technology, war, and human experi-ence. On the *Monitor,* men who wondered whether technology did not strip war of its pomp experienced the full flush of military glory. To the press, politicians, the public, and to the common soldier, the *Monitor* crew repre-sented the skill and mobility that seemed so lacking in a war descending into attrition.

Thus we find new significance in the otherwise well-known history of the *Monitor.* It is no longer the story of the heroic inventor and his impenetrable weapon thrusting themselves upon a doubtful and conservative bureaucracy. Rather it is a story of Ericsson and his associates working on multiple fronts to change the face, and the factories, of warfare. It is no longer the story of a heroic battle and the machine's epic loss soon after. Rather it is a story of peo-ple experiencing new machinery, attempting to make sense of its thrills, con-strictions, and politics, and sensing its power and impotence—both in glory and in frustration. Keeler's position, shared with much of the crew, was one of *contingent* heroism, always in question and always in need of maintenance. Women, encountered in public and in private, provided attention (and rewriting) that helped define the technology's success and maintain the crew's stature.

The public *Monitor* emerged in parallel with the ship itself. In the debates over naval warfare, symbolic exchange counted as much as technical exchange. Whether the *Monitor* won the battle is unknowable and irrelevant. It generated a *perception* of superiority (at least in the North), and that perception had material effects. Men with power became converted; they granted contracts for further ships of the same type. Foreign countries did the same, building copies or buying monitors abroad. Such actions, even if inspired by erroneous evaluations of the *Monitor*'s effectiveness, boosted the technology by extending its life and expanding the technical and operational envelope on which judgments of success could be based. The type was generally a failure: a few old monitors saw troubled service in Cuba in 1898, but the class essentially died out by the twentieth century.[46] Although other experiments had been inspired by the monitors, the dreadnought-era battleships of the early twentieth century shared little with the *Monitor* save rotating turrets (which they drew from Englishman Coles as much as from Ericsson). Nonetheless, the monitors had lasting impact. They sold the ideas that navies could build both ships and machines, that naval officers had to share their glory with designers and constructors, and that mechanical warfare, whatever its indignities, might also leave a place for human skill, and hence for heroism.

Let us return, then, to William Keeler, always a bit anxious in his buttons and uniform, sensing the tense relationship between war as experience and war as perception. Sometimes the two seemed harmonious: Keeler and the crew felt victorious and the public called them heroes. At other times the connections stretched: in a battle of appearances, the *Monitor* and the *Virginia* neutralized each other through threats, without actually clashing. At still other times, public images detached from lived experience: postwar struggles to claim the historical truth of the *Monitor* dislodged certainties memory had long ago solidified. Such was the paradox of the war machine: political imperatives of developing, selling, and deploying a complex technology gave new life to the heroic aspect of war by promoting the *Monitor* and its crew, even as the machine's operations reduced their status from warriors to operatives.

These relationships, more varied and dynamic than to permit a clean distinction between symbols and causes, open a new avenue for the study of technology and war. The intimacy of war's technical, social, and symbolic activities connects the *Monitor*'s history to a broader history of military technologies, one that awaits further examination by historians. Indeed, issues raised here appear in other arenas of technological history as well. What roles do individual people play in a world dominated by large technological systems? When is "resistance to change" a rational response to the social disruptions that accompany new technology? Do machines live up to their expectations and their public images? The *Monitor*'s history is a moment in the history of technology, the idea that machines are not independent entities but aggregates of materials, resources, skills, and symbols.

In December of 1862, while the *Monitor* was anchored off Newport News, Keeler traveled to Norfolk. There he was told that at any time during the month following the Battle of Hampton Roads, he and his comrades could have steamed to Norfolk and found the city and the *Virginia* paralyzed with fear, theirs for the taking. Keeler believed the rumors, and they burned inside him, for instead of steaming to an easy, clear victory, the *Monitor* had sat idle, missing the grand opportunity: "Everything we could wish for just within our grasp, honour, fame, notoriety, whatever the love of country or the desire to do a patriotic deed might prompt, to say nothing of the large amount of prize money involved—that all this just within our grasp should be lost & not by any fault of our own is vexatious, don't you think so?"[47]

Similar vexations, disquiets of uncertainty and loss, surround war machines in the twentieth century. Industrial war shocks into ambivalence modernity's credo of progress. Mechanized heroes recover skill and agency from faceless masses of military operatives. Spectacle joins maneuver and decision as determinants of battle's significance. Deterrence collapses the distinction between a weapon's appearance and its utility. Smart weapons displace heroism from the field to the laboratory, from warriors to engineers, and from spectacle to secrecy. Performance, both as action and as theater, resists measurement and defies closure. Irony and ambiguity, apparent opposites of industrial rationality, germinate in its shadows.

Notes

Frequently Cited Archival Sources

Allen collection	Gardiner Weld Allen Collection of Naval Pamphlets, Widener Library, Harvard University, Cambridge, Mass.
Bushnell papers	Cornelius Bushnell Papers, New York Historical Society, New York.
Ericsson papers, LC	John Ericsson Papers, Library of Congress, Washington, D.C.
Ericsson papers, NYHS	John Ericsson Papers, New York Historical Society, New York.
Fox papers	Gustavus Vasa Fox Papers, New York Historical Society, New York.
Geer letters	George S. Geer Letters, Mariner's Museum Library, Newport News, Virginia.
Goldsborough papers	Louis M. Goldsborough Papers, Library of Congress, Washington, D.C.
Keeler Letterbook	William F. Keeler Letterbook, Naval Academy Museum Library, Annapolis, Md.
Monitor log	Log of USS *Monitor,* National Archives, RG 45, Naval Records Collection of the Office of Naval Records and Library, Washington, D.C., vol. 161 PC-30, entry 392, subseries D-110.
NARA	Naval Records Collection of the Office of Naval Records and Library, Washington, D.C., RG 45.
ORN	Edward K. Rawson and Robert Wood, eds., *Official Records of the Union and Confederate Navies in the War of the Rebellion* (Washington, D.C., 1898).
Passaic papers	Papers of the U.S. Steamer *Passaic,* New York Public Library, New York.
Pierce papers	Frank H. Pierce Papers, New York Public Library, New York.
Welles papers, NYPL	Gideon Welles papers, New York Public Library, New York.

Introduction: A Strange Sort of Warfare

1. Keeler to his wife, Mar. 9, 1862, Keeler Letterbook. Keeler's letters were published in William F. Keeler, *Life aboard the U.S.S. Monitor, 1862: The Letters of Acting Paymaster William Frederick Keeler, U.S. Navy, to His Wife Anna,* ed. Robert Daly (Annapolis, Md., 1964); all Keeler letters will be cited from the original collection because of some omissions in published versions. For additional Keeler letters, see Robert W. Daly, ed., *Aboard the USS Florida: 1863–65: The Letters of Paymaster William Frederick Keeler, U.S. Navy, to His Wife, Anna* (Annapolis, Md., 1968).

2. Quoted in Keeler to his wife, Mar. 9, 1862, Keeler Letterbook. Almost exactly the same words are reported in Lt. Samuel Dana Greene's letter to his parents, Mar. 14, 1862, Naval Academy Museum Library, Annapolis, Md., and in G. B. Cannon, *Recollections of the* Monitor *and the* Merrimack *and Incidents of the Fights* (Burlington, Vt., 1875), in Allen collection. Cannon went aboard the *Monitor* with Fox.

3. Gideon Welles to John Worden, Mar. 15, 1862, in Edward K. Rawson and Robert Wood, eds., *Official Records of the Union and Confederate Navies in the War of the Rebellion* (Washington, D.C., 1898) (hereafter, ORN).

4. See the congressional commendations to the *Monitor* crew, Feb. 2, 1863, and July 11, 1862, ORN; Charles Boynton, *History of the Navy during the Rebellion* (New York, 1867); *New York Times,* Mar. 10, 1862. For more official histories, see James P. Delgado, "A Symbol of the People: Assessing the Significance of the U.S.S. Monitor," in J. Lee Cox, Jr., and Michael A. Jehle, eds., *Ironclad Intruder: U.S.S.* Monitor (Philadelphia, 1988), 37.

5. Henry Adams to Charles Francis Adams, Jr., Apr. 11, 1862, in *The Letters of Henry Adams,* vol. 1, *1858–1868* (Cambridge, Mass., 1982), 290.

6. Keeler to his wife, Feb. 13, 1862, Keeler Letterbook.

7. Eric Schmitt, "In Modern War, Fewer Heroes," *New York Times,* June 7, 1998.

8. Gerald Linderman, *Embattled Courage: The Experience of Combat in the American Civil War* (New York, 1987), 19.

9. William W. Blackford, *War Years with Jeb Stuart* (New York, 1945), 260–61, quoted in Linderman, *Embattled Courage,* 150.

10. James McPherson, *For Cause and Comrades: Why Men Fought in the Civil War* (New York, 1997), 100; Linderman, *Embattled Courage,* 7–16, details the fine distinctions between "courage," "duty," and "honor."

11. For struggles over mechanization in early industrial America, see Merritt Roe Smith, *Harper's Ferry Armory and the New Technology: The Challenge of Change* (Ithaca, N.Y., 1977); Judith A. McGaw, *Most Wonderful Machine: Mechanization and Social Change in Berkshire Paper Making, 1801–1885* (Princeton, N.J., 1987).

12. Nathaniel Hawthorne, "Chiefly about War Matters by a Peaceable Man," *Atlantic Monthly,* July 1862, 43.

13. The correspondence has been noted in Merton M. Sealts, Jr., *Melville's Reading* (Columbia, S.C., 1988), and in Leo B. Levy, "Hawthorne, Melville, and the Monitor," *American Literature* 37 (Mar. 1965): 33. See also Fredrick P. Kroeger, "Longfellow, Melville, and Hawthorne: The Passage into the Iron Age," *Illinois Quarterly* 33 (Dec. 1970): 37.

14. Herman Melville, "In the Turret" and "A Utilitarian View of the *Monitor*'s Fight," in Hennig Cohen, ed., *The Battle Pieces of Herman Melville* (New York, 1963), 55–57, 61–62. The quotation is from "A Utilitarian View of the *Monitor*'s Fight," 62.

15. Russel Blaine Nye, *Society and Culture in America, 1830–1860* (New York, 1974), 28;

Hugo A. Meier, "Technology and Democracy, 1800–1860," *Mississippi Valley Historical Review* 43, no. 4 (Mar. 1957): 618–40.

16. John Keegan, *The Face of Battle: A Study of Agincourt, Waterloo, and the Somme* (London, 1976), 32.

17. McPherson explicitly evokes Keegan, calling his book, *For Cause and Comrades,* "this Face of Battle for the Civil War," 184; Gerald Linderman, *Embattled Courage;* Bell Irvin Wiley, *The Life of Billy Yank: The Common Soldier of the Union* (Indianapolis, 1952); Bell Irvin Wiley, *The Life of Johnny Reb: The Common Soldier of the Confederacy* (Indianapolis, 1943); Dennis J. Ringle, *Life in Mr. Lincoln's Navy* (Annapolis, Md., 1998).

18. Judith McGaw, "The Experience of Early American Technology," in Judith McGaw, ed., *Early American Technology: Making and Doing Things from the Colonial Era to 1850* (Chapel Hill, N.C., 1994), 1–15.

19. Merritt Roe Smith and Leo Marx, eds. *Does Technology Drive History?: Essays on Technological Determinism* (Cambridge, Mass., 1994).

20. Donald MacKenzie, *Inventing Accuracy: A Historical Sociology of Nuclear Missile Guidance* (Cambridge, Mass., 1993), chap. 4.

21. Merritt Roe Smith, ed., *Military Enterprise and Technological Change: Perspectives on the American Experience* (Cambridge, Mass., 1985); Alex Roland, "Technology and War: The Historiographical Revolution of the 1980s," *Technology and Culture* 34, no. 1 (Jan. 1993): 117–34; Barton C. Hacker, "Military Institutions, Weapons, and Social Change: Toward a New History of Military Technology," *Technology and Culture* 35, no. 4 (Oct. 1994): 768–834; Barton C. Hacker, "Engineering a New Order: Military Institutions, Technical Education, and the Rise of the Industrial State," *Technology and Culture* 34, no. 1 (Jan. 1993): 1–27; Martin van Creveld, *Technology and War: From 2000 B.C. to the Present* (New York, 1989); William McNeill, *The Pursuit of Power: Technology, Armed Force, and Society from A.D. 1000 to the Present* (Chicago, 1982).

Chapter 1. Revising the Revolution, 1815–1861

1. Andrew Lambert, *Warrior: The World's First Ironclad, Then and Now* (London, 1987); David K. Brown, *Before the Ironclad* (London, 1990), chap. 14.

2. David K. Brown, "The Era of Uncertainty, 1863–1878," in Robert Gardiner, ed., *Steam, Steel, and Shellfire: The Steam Warship 1815–1905* (London, 1992), 75–94.

3. Isaac Newton, *The Monitor Ironclads* (Boston, 1864), 16.

4. *Times* (London), Mar. 25, 1862, 7; Henry Adams to Charles Francis Adams, Jr., Apr. 25, 1862, in *The Letters of Henry Adams,* vol. 1, *1858–1868* (Cambridge, Mass., 1982), 290; for other European press responses, see Robert Stanley McCordock, *The Yankee Cheesebox* (Philadelphia, 1938), chap. 17.

5. William C. Davis. *Duel between the First Ironclads* (Baton Rouge, 1975), 167.

6. Reported in the *Times* (London), Apr. 1, 1862, 6.

7. James Phinney Baxter, *The Introduction of the Ironclad Warship* (Cambridge, Mass., 1933).

8. Baxter, *Ironclad Warship,* 285, emphasis added.

9. James P. Delgado, "A Symbol of the People: Assessing the Significance of the U.S.S. Monitor," in J. Lee Cox, Jr., and Michael A. Jehle, eds., *Ironclad Intruder: U.S.S. Monitor* (Philadelphia, 1988), 37, 39; William N. Still, Jr. "The Historical Importance of the USS *Monitor,*" in William B. Cogar, ed., *Naval History: The Seventh Symposium of the U.S. Naval Academy* (Wilmington, Del., 1988), 79.

10. Michael Smith, "Selling the Moon: The U.S. Manned Space Program and the Triumph of Commodity Scientism," in Richard Wightman Fox and T. J. Jackson Lears, eds., *The Culture of Consumption: Critical Essays in American History, 1880–1980* (New York, 1983), 189; John W. Ward, "The Meaning of Lindbergh's Flight," *American Quarterly* 10 (1958): 3–16.

11. Leo Marx, "Technology: The Emergence of a Hazardous Concept," *Social Research* 64, no. 3 (fall 1997): 978.

12. For this distinction between two types of performance, see Robert C. Post, *High Performance: The Culture and Technology of Drag Racing* (Baltimore, 1994).

13. Brown, *Before the Ironclad,* 161. Not by coincidence, one of the chief architects of American nuclear strategy in the Cold War, Bernard Brodie, was trained as a historian of mid–nineteenth century naval technology. See Brodie, *Sea Power.*

14. Ken Alder, *Engineering the Revolution: Arms and Enlightenment in France, 1763–1815* (Princeton, N.J., 1997); John Fincham, *A History of Naval Architecture* (London, 1851), esp. the "Introduction to the history of shipbuilding, showing briefly the application of mathematical science to this art."

15. Carl von Clausewitz, *Principles of War* (1812), quoted in Daniel Pick, *War Machine: The Rationalization of Slaughter in the Modern Age* (New Haven, Conn., 1993), 36–37.

16. Brown, *Before the Ironclad,* chap. 6, argues that paddle wheels were not as vulnerable as they were reputed to be.

17. Ibid., 114, chaps. 9–10; Sir Howard Douglas, *On Naval Warfare with Steam* (London, 1860), 50–59; Andrew Lambert, "The Screw Propeller Warship," in Robert Gardiner, ed., *Steam, Steel, and Shellfire;* Basil Greenhill and Ann Giffard, *Steam, Politics, and Patronage: The Transformation of the Royal Navy 1815–54* (London, 1994).

18. John Dahlgren, *Shells and Shell Guns* (Philadelphia, 1856), 392, 395 (quotation on 392); Sir Howard Douglas, *Treatise on Naval Gunnery* (London, 1853), reprints Paixans's own response to Sinop and Paixans's conclusion that the future lay in shell guns (300–303); Brown, *Before the Ironclad,* 138; Andrew Lambert, *Battleships in Transition: The Creation of the Steam Battlefleet, 1815–1860* (London, 1984); Lambert, "The Screw Propeller Warship," 13.

19. Brown, *Before the Ironclad,* 157–58; H. W. Wilson, *Ironclads in Action: A Sketch of Naval Warfare from 1855 to 1895 with Some Account of the Development of the Battleship in England* (London, 1896), xxxi–xxxvi; Baxter, *Ironclad Warship,* 79. Lambert, *Battleships in Transition,* 92.

20. Robert Albion, *Forests & Sea Power: The Timber Problem of the Royal Navy: 1652-1862* (Cambridge, Mass., 1926), 406–7.

21. Brown, *Before the Ironclad,* chap. 8; Baxter, *Ironclad Warship,* 35–36, 90; Brodie, *Sea Power,* 136, 147.

22. Charles Hamley, "Fleets and Navies—France, Part I," *Blackwood's Magazine* 85 (June, Sept. 1859): 643–60; "Ironclad Ships of War," parts 1 and 2, *Blackwood's Magazine* 88 (Nov., Dec., 1860): 616–32, 633–49; Baxter, *Ironclad Warship,* 109.

23. Quoted in Brown, *Before the Ironclad,* 1, 174; Lambert, "The Screw Propeller Warship," 12.

24. Lambert, *Warrior,* 15–16; David McGee, "Floating Bodies, Naval Science: Science, Design and the *Captain* Controversy, 1860–1970," Ph.D. diss., University of Toronto, 1994, chap. 4.

25. Walter McDougall, *The Heavens and the Earth: A Political History of the Space Age* (New York, 1985), 134, 157.

26. One observer blamed the obscurity of the *Demologos* on the absence of a major battle: "The incalculable impulse given to steam as a factor in naval warfare . . . would have followed the success of the *Demologos* in battle, and which would have set forward the development of the times in this regard almost half a century." Frank M. Bennett, *The Steam Navy of the United States* (Pittsburgh, 1896), 9.

27. Harold Sprout and Margaret Sprout, *The Rise of American Naval Power 1776–1918* (Princeton, N.J., 1946), chap. 8.

28. K. Jack Bauer, "Naval Shipbuilding Programs, 1794–1860," *Miliary Affairs* 29, no. 1 (spring 1965): 29–40; Kurt Hackemer, "From Peace to War: United States Naval Procurement, Private Enterprise, and the Integration of New Technology, 1850–1865," (Ph.D. diss., Texas A&M University, 1994); Bennett, *Steam Navy of the United States,* chaps. 3–4. The Merrimack class also included the *Roanoake,* the *Colorado,* the *Wabash,* the *Minnesota,* and the *Niagara,* although the *Niagara* was a sloop, not a frigate. In addition to the *Merrimack,* the *Roanoake* and the *Minnesota* were in Hampton Roads during the battle.

29. Thomas R. Heinrich, *Ships for the Seven Seas: Philadelphia Shipbuilding in the Age of Industrial Capitalism* (Baltimore, 1997), 25.

30. Baxter, *Ironclad Warship,* 109.

31. Samuel Huntington, *The Soldier and the State* (Cambridge, Mass., 1957), 217.

32. Robert Schneller, Jr., *Quest for Glory: A Biography of Rear Admiral John A. Dahlgren* (Annapolis, Md., 1996). Spencer Tucker, *Arming the Fleet: U.S. Navy Ordnance in the Muzzle-Loading Era* (Annapolis, Md., 1989), chap. 6.

33. John H. Schroeder, "Matthew Calbraith Perry: Antebellum Precursor of the Steam Navy"; David K. Allison, "John A. Dahlgren: Innovator in Uniform"; William Stanton, "Matthew Fontaine Maury: Navy Science for the World"; Charles M. Todorich, "Franklin Buchannan: Symbol for Two Navies," all in James C. Bradford, ed., *Captains of the Old Steam Navy: Makers of the American Naval Tradition, 1840–1880* (Annapolis, Md., 1986); Alfred Thayer Mahan, *From Sail to Steam: Recollections of Naval Life* (New York, 1906).

34. Monte A. Calvert, *The Mechanical Engineer in America: Professional Cultures in Conflict* (Baltimore, 1967), chap. 13.

35. Bennett, *Steam Navy of the United States,* 2; Peter Karstens, *The Naval Aristocracy: The Golden Age of Annapolis and the Emergence of Modern American Navalism* (New York, 1972), 66–67. For a similar story in the British navy, see Greenhill and Giffard, *Steam, Politics, and Patronage,* 85–88.

36. Edward William Sloan III, *Benjamin Franklin Isherwood: Naval Engineer: The Years as Engineer in Chief, 1861–1869* (Annapolis, Md., 1965).

37. Benjamin Franklin Isherwood, *Experimental Researches in Steam Engineering* (Philadelphia, 1863), 327; Benjamin Franklin Isherwood, *Engineering Precedents for Steam Machinery* (New York, 1859).

38. Isherwood, *Experimental Researches in Steam Engineering,* xxv.

39. Sloan, *Isherwood;* Elting Morison, *Men, Machines and Modern Times* (Cambridge, Mass., 1966), chap. 6.

40. Bennett, *Steam Navy of the United States,* 185; Dean C. Allard, "Benjamin Franklin Isherwood: Father of the Modern Steam Navy," in Bradford, *Captains of the Old Steam Navy.*

41. William S. Dudley, *Going South: U.S. Navy Officer Resignations & Dismissals on the Eve of the Civil War* (Washington, D.C., 1981).

42. Gideon Welles, *Diary of Gideon Welles* (Boston, 1911), 41–54; Gideon Welles Papers, Library of Congress, Washington, D.C., reels 32-36, 37. Welles kept careful scrapbooks with editorials and press accounts of his conduct regarding the Norfolk yard.

43. James Russel Soley, "The Union and Confederate Navies," in Robert Underwood Johnson and Clarence Clough Buel, eds., *Battles and Leaders of the Civil War*, vol. 1, *From Sumter to Shiloh* (New York, 1887), 623. For Welles and the Department of the Navy during the Civil War, see Charles Oscar Paullin, "A Half Century of Naval Administration in America, 1861–1911: I—The Navy Department during the Civil War, 1861–1865," *United States Naval Institute Proceedings* 38 (Dec. 1912): 1309–36.

44. Gideon Welles, *Report of the Secretary of the Navy* (Washington, D.C., 1861), July 4, 1861, Dec. 2, 1861 (quotation on p. 14).

45. Ibid., July 4, 1861, 12.

46. *Philadelphia Examiner*, Mar. 21, 1861, quoted in Baxter, *Ironclad Warship*, 220. Richard Albion, *Makers of Naval Policy 1798–1947* (Annapolis, Md., 1980), 195.

47. "Act of Congress Authorizing the Construction of Iron-Clad Vessels," Aug. 3, 1861, printed in Gideon Welles, *Report of the Secretary of the Navy in Relation to Armored Vessels* (Washington, D.C., 1864), 1–2.

48. Welles, *Report of the Secretary of the Navy* (Washington, D.C., 1861), July 4, 1861.

49. The ad appeared in the *Baltimore Clipper*, the *Philadelphia Daily News*, the *Philadelphia Evening Journal*, the *Morning Courier & New York Enquirer*, the *New York Times*, and the *Baltimore Patriot*. NARA, AC, box 22, folder "Ads for Construction of Steam Vessels by Private Firms."

50. Advertisement for "Iron-Clad Steam Vessels," reprinted in Welles, *Report of the Secretary in Relation to Armored Vessels*, Dec. 2, 1861, 2.

51. "Report of Board to Examine Plans of Iron-Clad Vessels, under Act of August 3, 1861," in Welles, *Report of the Secretary in Relation to Armored Vessels*, Dec. 2, 1861, 3–7 (quotation on 4). In fact, when Welles appointed a similar board in January of 1862 to evaluate harbor and coast-defense ironclads, he tapped Engineer-in-Chief of the Navy Benjamin Isherwood, Edward Hartt, a naval constructor, and other experts in construction and engineering. The only member in common with the first committee was Smith, who, even after Hampton Roads, proclaimed his opposition to iron ships. Sloan, *Isherwood*, 57.

52. "Plans of Iron-Clad Vessels," 3.

53. On the *Galena*, see Hackemer, "From Peace to War," 213; on the *New Ironsides*, see "The U.S.S. Frigate *New Ironsides*," *Journal of the Franklin Institute* 53, no. 2 (Feb. 1867): 73–81; William H. Roberts, *USS* New Ironsides *in the Civil War* (Annapolis, Md., 1999).

54. "Plans of Iron-Clad Vessels."

55. John Ericsson, "The Building of the Monitor," in Johnson and Buel, *Battles and Leaders*, 731.

Chapter 2. Building a Ship, Speaking Success

1. Ericsson to Fox, Jan. 20, 1862, Ericsson papers, NYHS, emphasis added.

2. Ericsson to Arthur Hazelius, June 19, 1888, in William Conant Church, *The Life of John Ericsson* (New York, 1890), 2:308.

3. Samuel Smiles, *Lives of the Engineers* (1861; reprint, London, 1997).

4. William Conant Church, *The Life of John Ericsson* (New York, 1890), 1:1, 23. Church

says that Ericsson's papers before 1860 were destroyed at his request in 1886, as were his diaries. He was inspired to do it, Church says, by "the indiscretions of Carlyle's biographer." In looking over his diaries, Ericsson "found in [them]—as any man would under like circumstances—evidences of mistaken ideas and impressions that he did not care to perpetuate." *Life of Ericsson,* 2:307–8, 238. Also see Ruth White, *Yankee from Sweden: The Dream and the Reality in the Days of John Ericsson* (New York, 1960), a popular biography only slightly less celebratory than Church's.

5. Letter from John Ericsson, quoted in Church, *Life of Ericsson,* 1:17.

6. Church, *Life of Ericsson,* 1:41.

7. *Mechanic's Magazine,* Feb. 13, 1830, 434–35.

8. Church, *Life of Ericsson,* 1:48.

9. John Ericsson, "The Celebrated Trial of Locomotive Engines in the Liverpool and Manchester Railway," n.d., Ericsson papers, LC; *Mechanic's Magazine,* Oct. 17, 1829, 131–42 (quotation on 139); Oct. 24, 1829, 146–52.

10. John Ericsson, *Contributions to the Centennial Exhibition* (1876; reprint, Stockholm, 1976); Frank M. Bennett, *The Steam Navy of the United States* (Pittsburgh, 1896), 62-69.

11. Bernard Brodie, *Sea Power in the Machine Age* (Princeton, N.J., 1941), 36. Lee M. Pearson, "The Princeton and the Peacemaker: A Study in Nineteenth Century Research and Development Procedures," *Technology and Culture* 7 (Apr. 1966): 163–83.

12. Pearson, "Princeton and Peacemaker."

13. See letters from Ericsson to the navy concerning steam improvements, 1847, and proposals to the War Department for a steamer to navigate the Gulf of Mexico and Rio Grande, 1847–50, Ericsson papers, NYHS.

14. John Ericsson, "Caloric Ship Ericsson," n.d., Ericsson papers, NYHS; John M. Morrison, *History of American Steam Navigation* (New York, 1958), 427–29.

15. Church, *Life of Ericsson,* 1:92.

16. John Ericsson, "New System of Naval Attack," Sept. 26, 1854, copy in NARA, ser. AD, box 49, folder 1.

17. James Phinney Baxter, *The Introduction of the Ironclad Warship* (Cambridge, Mass., 1933), 184–85. Ericsson to Fox, Mar. 4, 1874, Fox papers, box 17.

18. Joseph Smith to J. Van Ness Phillip, May 8, 1861; Thomas Whitney to Smith, June 25, 1861; Cornelius Bushnell to Smith, July 14, 1861; Donald McKay to Smith, July 14, 1861, all in NARA, ser. AD, box 49, folder "Correspondence between Joseph Smith and Various Designers and Builders."

19. Richard Albion, *Makers of Naval Policy 1798–1947* (Annapolis, Md., 1980), 194–95. Senator Grimes's own account of the bill is quoted at length in William Salter, *The Life of James W. Grimes* (New York, 1876), 145–46; Bushnell to Welles, Mar. 9, 1877; Egbert Watson and Son interview with Cornelius Bushnell, in William S. Welles, ed., *The Story of the Monitor: The First Naval Conflict between Iron Clad Vessels* (1899; 2d ed., New Haven, Conn., 1906), 20–21. These accounts, though biased by the publisher and written many years after the events, were approved as accurate by Ericsson and Gideon Welles.

20. Ericsson to Bushnell, Aug. 7, Sept. 11, 1861 (quotation in Sept. 11 letter); Bushnell papers. NARA, subject file, "U.S. Navy, 1775–1910," ser. AD, and "Design & General Characteristics, U.S. Ships," box 49, folder 1, "USS *Monitor,* Contract Specifications, etc."; Bushnell papers.

21. Bushnell to Smith, Aug. 22, 24, 1861, Bushnell papers.

22. William N. Still, Jr., Monitor *Builders: A Historical Study of the Principal Firms and*

Individuals Involved in the Construction of USS Monitor (Washington, D.C., 1988), 8–9. Still argues that Corning functioned as a silent partner in the *Monitor* deal.

23. Watson and Son, interview with Bushnell; C. S. Bushnell, paper read before the Army and Navy Club of Connecticut, June 22, 1894, both in Welles, *The Story of the Monitor,* 21. Also see Whitney and Griswold to Smith, Sept. 21, 1861, NARA, ser. AD, box 49, "Specifications, etc."

24. Ericsson to Bushnell, Sept. 23, 1861, NARA, ser. AD, box 49.

25. Ericsson to James Gordon Bennett, Apr. 25, 1862, in Gideon Welles, *Report of the Secretary of the Navy in Relation to Armored Vessels* (Washington, D.C., 1864), 14. Ericsson to E. P. Dorr, Nov. 16, 1877, in Church, Life of *Ericsson,* 1:252-53.

26. John Ericsson, "Specification of an Impregnable Floating Battery, composed of Iron and Wood, Complete Ready for Service Excepting Guns, Ammunition, and Stores," NARA, subject file, "1775–1910," ser. AD, "Design & General Characteristics, U.S. Ships," box 49, folder 1.

27. Ibid.

28. Ericsson letter, Nov. 16, 1877, in Church, *Life of Ericsson,* 1:252-53.

29. Smith to Ericsson, Sept. 21, 1861, Ericsson papers, NYHS.

30. "Report of Board to Examine Plans of Iron-clad Vessels, under Act of August 3, 1861," Sept. 16, 1861, in Welles, *Report of the Secretary in Relation to Armored Vessels,* Dec. 2, 1861, 3.

31. Kurt Hackemer, "From Peace to War: United States Naval Procurement, Private Enterprise, and the Integration of New Technology, 1850–1865," Ph.D. diss., Texas A&M University, 1994, 59.

32. Ericsson to Smith, Sept. 28, 1861, NARA, subject file, "U.S. Navy, 1775–1910," ser. AD, "Design & General Characteristics, U.S. Ships," box 49, folder 1. Also see Bushnell to Smith, Sept. 28, 1861, NARA, subject file, "U.S. Navy, 1775–1910," ser. AD, "Design & General Characteristics, U.S. Ships," box 49, folder "Correspondence between Joseph Smith and Various Designers and Builders." Winslow's letter to Ericsson, Nov. 2, 1862, seems to contradict this point, because Winslow read of the guarantee in *Scientific American* and wrote, "This is news to me." Ericsson papers, NYHS.

33. Ericsson to Smith, Oct. 2, 1861, NARA, subject file, "U.S. Navy, 1775–1910," ser. AD, "Design & General Characteristics, U.S. Ships," box 49, folder 1.

34. A good technical description of the *Monitor* can be found in Donald Canney, *The Old Steam Navy,* vol. 2, *The Ironclads, 1842-1885* (Annapolis, Md., 1993), 25–31; Ernest W. Peterkin, ed., "Drawings of the USS *Monitor,*" USS *Monitor* Historical Report series, U.S. Dept. of Commerce, National Oceanic and Atmospheric Administration, Washington, D.C., vol. 1, no. 1, Dec. 1985. For detailed specifications of the engine, see Benjamin Franklin Isherwood, *Experimental Researches in Steam Engineering* (Philadelphia, 1863), 327.

35. Some evidence indicates that a periscope-like device was designed into the turret so the crews could see out. *Scientific American* reported, "The turret is pierced in different places with four holes for the insertion of telescopes and just reflectors are fixed to bend the rays of light which comes in a direction parallel with the guns through the axis of the telescope, which is crossed by a vertical thread of spider's web though the line of collumation . . . [leading to the capability to fire with] unprecedented accuracy. *Scientific American,* Mar. 22, 1862, 177. Also, Ericsson's letter to Napoleon III, proposing his "new system of naval attack," mentions, "Reflecting telescopes capable of being protruded or

withdrawn at pleasure also afford a distant view of surrounding objects." Ericsson to Napoleon III, 1854, Ericsson papers, NYHS. No evidence from the *Monitor's* operations suggests that these telescopes were actually installed, possibly because of the hasty construction.

36. "Agreement between Rowland, Winslow, Griswold, and Bushnell," Oct. 25, 1861. NARA, AD, "Design & General Characteristics, U.S. Ships," box 49, folder 1.

37. Rowland to Bushnell, Nov. 11, 1861. "Contract between Bushnell and Franics Clusk for a building in Rowland's Yard." n.d. [probably Oct. 1861], Bushnell papers.

38. Bushnell to Smith, Oct. 21, 1861, NARA, subject file, "U.S. Navy, 1775–1910," ser. AD, "Design & General Characteristics, U.S. Ships," box 49; Peterkin, "Drawings of the USS Monitor."

39. Still, Monitor *Builders,* 23–26; NARA, subject file, "U.S. Navy, 1775–1910," ser. AD, "Design & General Characteristics, U.S. Ships," box 49. On locomotive construction, which has similarities to iron ship construction, see John K. Brown, *The Baldwin Locomotive Works 1831–1915* (Baltimore, 1995).

40. Robert Stanley McCordock, *The Yankee Cheesebox* (Philadelphia, 1938), 44–45.

41. Smith to Ericsson, Mar. 3, 1853, Ericsson papers, NYHS.

42. Ericsson to Smith, Oct. 24, 1861, NARA, subject file, "U.S. Navy, 1775–1910," ser. AD, "Design & General Characteristics, U.S. Ships," box 49, folder "Correspondence between Joseph Smith and Various Designers and Builders."

43. Ericsson to Smith, Oct. 14, 17, 1861, NARA, AD, "Design & General Characteristics, U.S. Ships," box 49, folder 1.

44. Smith to Ericsson, Oct. 2, 1861, Ericsson papers, NYHS.

45. Ibid., Oct. 19, 1861. Ericsson to Smith, Oct. 18, 1861, NARA, RG 45, folder "Correspondence between Joseph Smith and Various Designers and Builders."

46. Ericsson to Smith, Oct. 25, 1861, NARA, AD, "Design & General Characteristics, U.S. Ships," box 49, folder 1.

47. Earl J. Hess, "Northern Response to the Ironclad: A Prospect for the Study of Military History," *Civil War History* 31 (June 1985), 140.

48. Corning to Welles, Dec. 25, 1861, quoted in Baxter, *Ironclad Warship,* 278.

49. Winslow to Ericsson, Jan. 10, 1862, in Church, *Life of Ericsson,* 1:277–78.

50. Ibid., quoted in Church, *Life of Ericsson,* 2:2.

51. Griswold to Ericsson, Jan. 24, 1862, Ericsson papers, NYHS.

52. Ibid., Feb. 13, 1862, quoted in Baxter, *Ironclad Warship,* 279. Also see Winslow to Ericsson, Jan. 6, 1862, quoted in Baxter, *Ironclad Warship,* 279. For Baxter, this episode shows that the Department of the Navy was convinced of turreted ironclads even before Hampton Roads and was merely deciding between the two systems of turrets, Ericsson's and Coles's (283). This judgment is confirmed by the fact that on Feb. 20, 1862, the navy advertised for "Iron-Clad steamers" on which "guns must train on all points of comps without change in vessel's position." NARA, AC, "Construction of U.S. Ships," 0-1859, "Misc. Material, 1857–61," box 22, folder "USN Ads for Construction of Steam Vessels by Private Firms." One undated version of the *Monitor* specification even refers to the turret as a "Cole tower." "The Monitor," handwritten specification, NARA, subject file, "U.S. Navy, 1775–1910," ser. AD, "Design & General Characteristics, U.S. Ships," box 49, folder 1.

53. Barnes to Ericsson, Feb. 21, 1862, Ericsson papers, NYHS.

54. Bushnell to Ericsson, Feb. 26, 1862, Ericsson papers, NYHS.

55. Winslow to Ericsson, Mar. 3, 1862, Ericsson papers, NYHS.

56. Smith to Ericsson, Jan. 14, Feb. 2, 1862, Ericsson papers, NYHS.

57. Fox to Ericsson, Feb. 21, 1862, Ericsson papers, NYHS.

58. Griswold to Ericsson, Feb. 24, 1862, Ericsson papers, NYHS.

59. William N. Still, Jr., *Ironclad Captains: The Commanding Officers of the USS "Monitor"* (Washington, D.C., 1988), 4–5.

60. White to Frank H. Pierce, Sept. 18, 1886, Pierce papers.

61. USS *Monitor* muster roll, RG 24, National Archives, Washington, D.C.; Irwin M. Berent, "The Crewmen of the USS Monitor: A Biographical Directory," USS *Monitor* National Marine Sanctuary, Historical Report series, U.S. Dept. of Commerce, National Oceanographic and Atmospheric Administration, Washington, D.C.; for naval personnel in general, see Dennis J. Ringle, *Life in Mr. Lincoln's Navy* (Annapolis, Md., 1998), chaps. 2-3.

62. Paymaster's Letterbook, *Passaic* papers.

63. Durst, Butts, and Driscoll, to Frank H. Pierce, May 28, 1885, Pierce papers; Geer letters.

64. Ericsson to Smith, Oct. 4, 1861, NARA, subject file, "U.S. Navy, 1775–1910," ser. AD, "Design & General Characteristics, U.S. Ships," box 49; Smith to Stimers, Nov. 5, 1861, in Julia Stimers Dubrow, ed., *The "Monitor" and Alban B. Stimers* (Orlando, Fla., 1936).

65. Monte A. Calvert, *The Mechanical Engineer in America: Professional Cultures in Conflict* (Baltimore, 1967), 252.

66. Stimers to Fox, Feb. 3, 1862, Fox papers.

67. Monitor log, Feb. 5–Mar. 3, 1862.

Chapter 3. William Keeler's Epistolary *Monitor*

1. Keeler to his wife, Jan. 12, 1862, Keeler Letterbook.

2. Ibid., Feb. 9, 1862.

3. Ibid.

4. Ibid., Jan. 12, Feb. 9, 1862.

5. Ibid., Feb. 28, 1862.

6. R. G. Dun and Company Collection, Illinois Volume 114, Connecticut Volume 2, Baker Library, Harvard Business School.

7. Robert Daly, introduction to William F. Keeler, *Life aboard the U.S.S. Monitor, 1862: The Letters of Acting Paymaster William Fredrick Keeler, U.S. Navy, to His Wife Anna,* ed. Robert Daly (Annapolis, Md., 1964), xiv. Daly states that Keeler was born in Utica, New York, but Keeler wrote in a letter to Frank H. Pierce, May 27, 1885, that he had been born in New York City and spent his first thirteen years there. Pierce papers.

8. Keeler to his wife, Feb. 22, June 30, 1862, Keeler Letterbook.

9. Dunn and Company Collection, Illinois Volume 114.

10. Keeler to Pierce, May 27, 1885, Pierce papers; Keeler to his wife, Feb. 13, 1862, Keeler Letterbook.

11. Keeler to his wife, Apr. 25, May 12, 1862, Keeler Letterbook.

12. Ibid., Feb. 9, 1862.

13. These conclusions are based on the papers and account books of the steamer USS *Passaic, Passiac* papers.

14. Keeler to his wife, Apr. 25, 1862, Keeler Letterbook.

15. Geer to his wife, Mar. 18, 16, 1862, Geer letters.

16. Keeler to his wife, Mar. 26, 1862, Keeler Letterbook.

17. Ibid., Feb. 9, 1862.

18. Geer to his wife, June 5, 1862, Geer letters.

19. James McPherson, *For Cause and Comrades: Why Men Fought in the Civil War* (New York, 1997), 12.

20. Keeler to his wife, Apr. 25, 1862, Keeler Letterbook.

21. Ibid., Mar. 16, 26, 1862.

22. Geer to his wife, Mar. 24, 1862, Geer letters.

23. Keeler to his wife, Apr. 25, May 1, 1862, Keeler Letterbook.

24. Ibid., June 14, 1862. See also Geer to his wife, May 26, June 6, when he received similar orders. Geer letters.

25. Keeler to his wife, Feb. 28, Mar. 5, 1862, Keeler Letterbook.

26. Ibid, Feb. 13, 25, Mar. 4, 1862.

27. Ibid., Mar. 5, 1862.

28. Ibid., Mar. 11, 1862.

29. Ibid., Feb. 28, 1862.

30. Ibid., May 5, 6, 1862; Anna E. Keeler to Frank H. Pierce, Dec. 4, 1883, May 23, 1886, Pierce papers; Geer on burning his letters, Geer to his wife, Mar. 18, 1862, Geer letters.

31. Keeler to his wife, Feb. 9, 1862, Keeler Letterbook.

32. Ibid., June 30, 1862.

33. Ibid., Mar. 9, 18, May 1, 1862.

34. *New York Times,* Mar. 30, 1862; Keeler to his wife, Apr. 3, 1862, Keeler Letterbook.

35. Keeler to his wife, Apr. 8, 1862, Keeler Letterbook.

36. Ibid., Apr. 25, 1862.

37. Ibid., Mar. 18, 1862.

38. Ibid., Mar. 29, 1862.

39. Margaret Creighton and Lisa Norling, introduction to Margaret Creighton and Lisa Norling, eds., *Iron Men, Wooden Women: Gender and Seafaring in the Atlantic World 1700–1920* (Baltimore, 1996), xiii; Margaret Creighton, *Rites and Passages: The Experience of American Whaling, 1830–1870* (Cambridge, 1995), chap. 7, "Sailors, Sweethearts, and Wives: Gender and Sex in the Deepwater Workplace."

40. Keeler to his wife, May 6, 1862, Keeler Letterbook; Gerald Linderman, *Embattled Courage: The Experience of Combat in the American Civil War* (New York, 1987), 8.

41. Keeler to Pierce, May 27, 1885, Pierce papers.

Chapter 4. Life in the Artificial World

1. *Monitor* log, Mar. 6, 1862; John Driscoll to Franklin Delano Roosevelt, 1926, NARA, ser. HA, box 174.

2. Keeler to his wife, May 6, 1862, Keeler Letterbook.

3. Rosalind Williams, *Notes on the Underground: An Essay on Technology, Society, and the Imagination* (Cambridge, Mass., 1990), 8.

4. Keeler to his wife, Mar. 5, 1862, Keeler Letterbook.

5. Ibid., Mar. 13, 1862; Robert Erwin Johnson, *Rear Admiral John Rodgers 1812-1882* (Annapolis, Md., 1967), 227. Alva Hunter, on the monitor *Nahant,* noted, "Waves flowing over [the skylights] caused very curious and interesting lighting effects in the rooms beneath. The water twisted about and boiled in the deadlight-cavity, and, being more or

less charged with air bubbles, the constantly changing shades of green light were a pleasure to study." *A Year on a "Monitor" and the Destruction of Fort Sumter,* ed. Craig L.
Symonds (Columbia, S.C., 1987), 18.

6. Keeler to his wife, Dec. 6, 1862, Keeler Letterbook.

7. Ibid.

8. Geer to his wife, Mar. 15, 1862, Geer letters.

9. Nathaniel Hawthorne, "Chiefly about War Matters by a Peaceable Man," *Atlantic
Monthly,* July 1862, 58.

10. Margaret Creighton notes a similar fact aboard whaling ships, where the absence
of women meant that at least some men had to perform traditionally feminine domestic
duties. *Rites and Passages: The Experience of American Whaling, 1830–1870* (Cambridge,
1995), 185.

11. *Times* (London), Mar. 27, 1862, 9.

12. Jules Verne, *Twenty Thousand Leagues under the Sea,* trans. Walter James Miller (New
York, 1976), 9.

13. Keeler to his wife, Mar. 5, 1862, Keeler Letterbook; the *Monitor* log, Mar. 3, 1862,
reports the steward's dismissal.

14. Keeler to his wife, Aug. 18, 1862, Keeler Letterbook.

15. Hess notes this tendency in public response to the *Monitor.* Earl J. Hess, "Northern
Response to the Ironclad: A Prospect for the Study of Military History," *Civil War History* 31 (June 1985): 140.

16. Frank Butts, *My First Cruise at Sea and the Loss of the Ironclad "Monitor"* (Providence, R.I., 1878); Geer to his wife, Apr. 21, June 4, 1862, Geer letters.

17. Geer to his wife, Mar. 24, 1862, Geer letters.

18. William Jeffers to Gideon Welles, May 16, 1862, in Edward K. Rawson and Robert
Wood, eds., *Official Records of the Union and Confederate Navies in the War of the Rebellion* (Washington, D.C., 1898); Keeler to his wife, Aug. 15, 1862, Keeler Letterbook.

19. Keeler to his wife, May 23, 1862, Keeler Letterbook.

20. Jeffers to Goldsborough, May 22, 1862, in Gideon Welles, *Report of the Secretary of
the Navy in Relation to Armored Vessels* (Washington, D.C., 1864), 27–29.

21. Keeler to his wife, June 2, Mar. 30, Apr. 21, 1862, Keeler Letterbook.

22. Ericsson to Bourne, Jan. 15, 1865, quoted in William Conant Church, *The Life of
John Ericsson* (New York, 1890), 2:68.

23. Hawthorne "Chiefly about War Matters," 58.

24. Greene to his parents, Mar. 14, 1862, Naval Academy Museum Library, Annapolis,
Md.

25. Albert Campbell to his wife, Mar. 10, 1862, copy in Pierce papers.

26. Keeler to his wife, Mar. 5, 1862, Keeler Letterbook.

27. Alban B. Stimers to his wife, Mar. 8, 1862; Stimers to his father, May 5, 1862, in Julia
Stimers Dubrow, ed., *The "Monitor" and Alban B. Stimers* (Orlando, Fla., 1936); John Worden to Gideon Welles, Jan. 5, 1862, Welles papers, NYPL; John M. White [John Driscoll]
to Pierce, Sept. 18, 1886, Pierce papers.

28. Greene to his parents, Mar. 10, 1862, Naval Academy Museum Library, Annapolis,
Md.

29. Keeler to his wife, Mar. 6, 1862, Keeler Letterbook.

30. Stimers to his wife, Mar. 8, 1862, in Stimers Dubrow, *The "Monitor" and Alban B.
Stimers;* White to Pierce, Sept. 18, 1886, Pierce papers.

31. Stimers to Ericsson, Mar. 9, 1862, Ericsson papers, NYHS; Stimers to his wife, Mar. 8, 1862, in which he wrote, "I never saw a vessel more buoyant or less shocked in a heavy sea way, than she was yesterday."

32. Ericsson to Stimers, Mar. 13, 1862, in Stimers Dubrow, *The "Monitor" and Alban B. Stimers.*

33. The increase in height of the air intakes to keep them above water when the vessel penetrated the waves demonstrates an early form of the periscope/snorkel of modern submarines. In later monitors, Ericsson moved the forward pilothouse to a position atop the gun turret, making an early conning tower.

34. Keeler to his wife, May 6, 1862, Keeler Letterbook.

35. Ibid., Mar. 9, 1862.

36. John Worden, "The Monitor's First Trip," *Youth's Companion,* Aug. 15, 1895; Pierce papers.

37. Hawthorne, "Chiefly about War Matters," 57.

38. Greene to his parents, Mar. 14, 1862, Naval Academy Museum Library, Annapolis, Md.

Chapter 5. The Battle of Hampton Roads

1. John Sherman Long, "The Gosport Affair, 1861," *Journal of Southern History* 23 (May 1957): 155–72; William N. Still, Jr., *Iron Afloat: The Story of the Confederate Armorclads* (Nashville, Tenn., 1971); William N. Still, Jr., *Confederate Shipbuilding* (Columbia, S.C., 1969).

2. Thomas Oliver Selfridge, Jr., *What Finer Tradition: The Memories of Thomas O. Selfridge, Jr, Rear Admiral, U.S.N.* (Columbia, S.C., 1987), 47.

3. Robert Bruce, *Lincoln and the Tools of War* (1954; reprint, Chicago, 1989).

4. Gideon Welles, *Diary of Gideon Welles,* vol. 1 (Boston, 1911), 62-66 (quotation on 66).

5. Greene to his parents, Mar. 10, 1862, Naval Academy Museum Library, Annapolis, Md.

6. McClellan to Wool, Mar. 9, 1862, in Edward K. Rawson and Robert Wood, eds., *Official Records of the Union and Confederate Navies in the War of the Rebellion* (Washington, D.C., 1898) (hereafter, ORN), ser. 1, 7:74–75.

7. For a good account of the strategic setting of Hampton Roads, see Robert W. Daly, *How the Merrimack Won: The Strategic Story of CSS Virginia* (New York, 1957); "Tales of Old Fort Monroe," nos. 6, 14, Fort Monroe Casemate Museum, Fort Monroe, Va.

8. Keeler to his wife, Apr. 11, 1862, Keeler Letterbook. William N. Still, Jr., "The Historical Significance of the U.S.S. Monitor," in J. Lee Cox, Jr., and Michael A. Jehle, eds., *Ironclad Intruder: U.S.S.* Monitor (Philadelphia, 1988).

9. Keeler to his wife, Mar. 9, 1862, Keeler Letterbook.

10. G. J. Van Brunt to Welles, Mar. 10, 1862, in Gideon Welles, *Report of the Secretary of the Navy in Relation to Armored Vessels* (Washington, D.C., 1864), 17–19 (quotation on 18).

11. Greene to his parents, Mar. 14, 1862, Naval Academy Museum Library, Annapolis, Md.

12. See letters in the Pierce papers from William Durst, Frank Butts, William Keeler, Louis Stodder, Samuel Taylor, Joseph Watters, John M. White, 1885–86, surrounding the publication of the *Century* magazine collection. See discussion in the conclusion.

13. The *Raleigh Standard,* Mar. 10, 1862, in Frank Moore, ed., *The Rebellion Record: A*

Diary of American Events, vol. 4 (New York, 1862), 276. For a collection of accounts of the battle from numerous vantage points, see the Allen collection. Accounts of Confederate victory include the pamphlets John B. Piet, *The Story of the Confederate States Ship, "Virginia," (Once Merrimack) Her Victory of the* Monitor (Baltimore, 1879); and E. V. White, *The First Iron-Clad Naval Engagement in the World* (Portsmouth, Va., 1906), both in Allen collection; Daly, *How the Merrimack Won.* For official reports, see U.S. War Department, *The War of the Rebellion: A Compilation of the Official Records of the Union and Confederate Armies,* 128 vols. (Washington, D.C., 1885–1902), ser. 1, 5:30–51; for more press accounts, see Robert Stanley McCordock, *The Yankee Cheesebox* (Philadelphia, 1938), chap. 9.

14. Hunter Davidson, an officer aboard the *Merrimack,* quoted in the *New York Sun,* May 9, 1894; Jefferson Davis, Commendation of the *Virginia* Crew to the Senate and House of Representatives of the Confederate States of America, Mar. 11, 1862. Moore, *Rebellion Record,* 4:271.

15. Stimers to his father, Mar. 11, 1862, in Julia Stimers Dubrow, ed., *The "Monitor" and Alban B. Stimers* (Orlando, Fla., 1936).

16. Daniel C. Logue to Welles, Mar. 11, 1862, ORN, ser. 1, 7:25.

17. Greene, "In the *Monitor* Turret," in Robert Underwood Johnson and Clarence Clough Buel, eds., *Battles and Leaders of the Civil War,* vol. 1, *From Sumter to Shiloh* (New York, 1887), 725. Greene wrote this account years after the war while defending his own performance and the *Monitor's* poor shooting, so it should be evaluated with a critical eye. Nevertheless, it jibes with others' accounts, including Keeler's and Worden's, and with Greene's own letter to his parents of Mar. 14, 1862.

18. Keeler to Pierce, Nov. 15, 1885, Pierce papers.

19. Greene, "In the *Monitor* Turret," 723–24.

20. Keeler to his wife, May 6, 1862, Keeler Letterbook.

21. Quoted in ibid., Mar. 13, 1862. Also see John Worden to Gideon Welles, Jan. 5, 1868, Welles papers, NYPL.

22. *Baltimore American,* Mar. 13, 1862, in Moore, *Rebellion Record,* 276.

23. Greene to his parents, Mar. 14, 1862, Naval Academy Museum Library, Annapolis, Md.

24. Flye to Pierce, Mar. 17, 1886, Pierce papers.

25. Keeler to his wife, Mar. 13, 1862, Keeler Letterbook.

26. Grimes speech before Congress, Mar. 13, 1862, quoted in William Salter, *The Life of James W. Grimes* (New York, 1876), 180–81.

27. James P. Delgado, "A Symbol of the People, Assessing the Significance of the U.S.S. Monitor," in Cox and Jehle, *Ironclad Intruder;* McCordock, *The Yankee Cheesebox,* chap. 9. Earl Hess discusses the contrast between the euphoric response to the *Monitor* and the more pragmatic response to ironclads on western rivers. "Northern Response to the Ironclad: A Prospect for the Study of Military History," *Civil War History* 31 (June 1985): 140.

28. Keeler to his wife, Mar. 13, 18, 1862, Keeler Letterbook; Geer to his wife, Mar. 10, 26, 1862, Geer letters.

29. Keeler to his wife, Mar. 28, 1862, Keeler Letterbook.

30. Ibid., May 7, 1862.

31. Ibid., Mar. 31, 1862.

32. Ibid.

33. Curtis to Charles Eliot Norton, June 26, 1862, quoted in Edward M. Clay, *George W. Curtis* (Boston, 1894), 156; also quoted in Stanton Garner, *The Civil War World of Her-*

man Melville (Lawrence, Kans., 1993), 154. With "Chiefly about War Matters," Hawthorne began as a regular contributor to the magazine; everything he subsequently published first appeared in the *Atlantic*. Edward Mather, *Nathaniel Hawthorne: A Modest Man* (1940; reprint, Westport, Conn., 1970), 325. The description of Lincoln in Hawthorne's article ("I should have taken him for a country schoolmaster as soon as anything else . . . this sallow, queer, sagacious visage,") was so frank it bordered on disrespect, forcing James Fields to delete it from the published version. In its stead, Hawthorne added footnotes that commented on the text "in the style of a shocked and patriotic Aunt." Readers were angered that the article had been cut and objected to the impertinence of the footnotes, not realizing that Hawthorne himself was mocking the *Atlantic's* "pompous patriotism." The original article, including the description of Lincoln, appears in *The Complete Works of Nathaniel Hawthorne* (Boston, 1882), 12:299–345.

34. Leo Marx, *The Machine in the Garden: Technology and the Pastoral Ideal in America* (London, 1964), 276–77; Taylor Stoehr, *Hawthorne's Mad Scientists: Pesudo Sciences and Social Sciences in Nineteenth Century Life and Letters* (Hamden, Conn., 1978).

35. Marx, *The Machine in the Garden,* 247.

36. Nathaniel Hawthorne, *The American Notebooks,* ed. Randall Stewart (New Haven, Conn., 1932), 102, quoted in Marx, *The Machine in the Garden,* 13.

37. Horatio Bridge, *Personal Recollections of Nathaniel Hawthorne* (New York, 1893), 169, quoted in Randall Stewart, "Hawthorne and the Civil War," *Studies in Philology* 34 (1937): 97.

38. Newton Arvin, *Nathaniel Hawthorne* (Boston, 1929), 269–70; Stewart, "Hawthorne and the Civil War," shows that Hawthorne's early feelings on the war found their way, "*mutatis mutandis,*" into the novel *Septimius Felton* (92–93).

39. Nathaniel Hawthorne, "Chiefly about War Matters by a Peaceable Man," *Atlantic Monthly,* July 1862, 50.

40. Ibid., 51.

41. Stewart, "Hawthorne and the Civil War," 95.

42. Hawthorne "Chiefly about War Matters," 57.

43. Ibid.

44. Ibid., 58.

45. Ibid., 59.

46. Keeler to his wife, Mar. 11, 1862, Keeler Letterbook.

47. Campbell to his wife, Mar. 10, 1862, copy in Pierce papers.

48. Keeler to his wife, Mar. 18, Apr. 15, 1862, Keeler Letterbook; see also Geer to his wife, Mar. 24, 1862, Geer letters.

Chapter 6. Iron Ship in a Glass Case, April–September 1862

1. Several accounts that follow this form are William C. Davis, *Duel between the First Ironclads* (Baton Rouge, 1975); A. A. Hoehling, *Thunder at Hampton Roads* (New York, 1993); Edward M. Miller, *USS "Monitor:" The Ship That Launched a Modern Navy* (Annapolis, Md., 1978); James Tertius deKay, *"Monitor"* (New York, 1997); and the collection of firsthand accounts in Robert Underwood Johnson and Clarence Clough Buel, eds., *Battles and Leaders of the Civil War,* vol. 1, *From Sumter to Shiloh* (New York, 1887).

2. Keeler to his wife, Apr. 15, 1862, Keeler Letterbook.

3. Ericsson to Fox, Mar. 15, 19, 1862, Fox papers.

4. Ericsson papers, NYHS. The Library of Congress has a complete schedule of the

Monitor's finances and estimates a total cost of $195,000, leaving a profit of $80,000, of which Ericsson received one-quarter plus a $1,000 fee for services, for a total of $21,000. William Conant Church, *The Life of John Ericsson* (New York, 1890), 1:270, reprints this schedule.

5. Smith to Ericsson, Mar. 17, 1862, Ericsson papers, NYHS. These six monitors would be the *Passaic,* the *Montauk,* the *Catskill,* the *Patapsco,* the *Lehigh,* and the *Sangamon.*

6. William N. Still, Jr., "The American Civil War," in Robert Gardiner, ed., *Steam, Steel, and Shellfire: The Steam Warship 1815–1905* (London, 1992), 67, and Donald Canney, *The Old Steam Navy,* vol. 2, *The Ironclads, 1842–1885* (Annapolis, Md., 1993), have charts and surveys of the complete monitor program.

7. Fox to Ericsson, dated Mar. 8, 1862 (the date is incorrect, though the letter was probably written sometime in March), Fox papers. For the crew's reports of the vessel leaking at anchor, see Geer to his wife, Apr. 21, June 4, 1862, Geer letters.

8. Fox to Ericsson, Apr. 15, 1862, Ericsson papers, NYHS.

9. Newton to Ericsson, May 7, 1862, Ericsson papers, NYHS.

10. Ericsson to Stimers, Mar. 13, 1862, in Julia Stimers Dubrow, ed., *The "Monitor" and Alban B. Stimers* (Orlando, Fla., 1936). Ericsson to Fox, Mar. 13, 14, 15, 1862, Fox papers.

11. Fox to Goldsborough, Mar. 8, 1862, in Edward K. Rawson and Robert Wood, eds., *Official Records of the Union and Confederate Navies in the War of the Rebellion* (Washington, D.C., 1898) (hereafter, ORN). On Jeffers as the possible first choice for captain, see Smith to Stimers, Jan. 15, 1862, in Stimers Dubrow, *The "Monitor" and Alban B. Stimers.*

12. Keeler to his wife, June 23, 1862, Keeler Letterbook; Geer to his wife, Mar. 26, Apr. 21, June 5, 1862, Geer letters.

13. Geer to his wife, Mar. 15, Apr. 7, 1862, Geer letters.

14. Keeler to his wife, Mar. 18, 1862, Keeler Letterbook.

15. Ibid., Aug. 19, 1862; statement by Flag Officer Tattnal, C.S.N., Apr. 8, 1862, ORN.

16. Keeler to his wife, Mar. 30, 1862, Keeler Letterbook.

17. Gideon Welles to Gustavus Fox, Mar. 10, 1862, ORN; William N. Still, Jr., *Ironclad Captains: The Commanding Officers of the USS "Monitor"* (Washington, D.C., 1988), 41.

18. Gustavus Fox, telegraph to Postmaster General M. Blair, Mar. 10, 1862, ORN. For the Union commanders' caution in their use of the *Monitor,* see James Phinney Baxter, *The Introduction of the Ironclad Warship* (Cambridge, Mass., 1933), 298.

19. Goldsborough to his wife, Apr. 14, 15, 30, 1862; Welles to Goldsborough, Apr. 7, 1862, Goldsborough papers.

20. Keeler to his wife, Apr. 11, 1862, Keeler Letterbook; Wool to Stanton, Apr. 11, 1862; Goldsborough to Welles, Apr. 12, 1862, ORN; William N. Still, Jr., "I Will Fight on My Own Terms," *Civil War Times Illustrated* 17, no. 10 (Feb. 1979): 12-20.

21. Keeler to his wife, Apr. 11, 1862, Keeler Letterbook.

22. Goldsborough to his wife, May 6, 1862, Goldsborough papers.

23. Keeler to his wife, May 4, 1862, Keeler Letterbook.

24. Ibid.

25. Geer to his wife, Mar. 18, 24, Apr. 13, 16, 1862, Geer letters.

26. Keeler to his wife, Mar. 30, 1862, Keeler Letterbook.

27. Stimers to Fox, Apr. 14, 1862, Fox papers.

28. Newton to Ericsson, May 7, 1862, Ericsson papers, NYHS; Geer to his wife, Apr. 6, 1862, Geer letters.

29. "The Monitor Boys" to Worden, Apr. 24, 1862, in Johnson and Buel, *Battles and Leaders,* 726.

30. Keeler to his wife, Mar. 30, 1862, Keeler Letterbook.

31. Ibid., May 7, 1862.

32. Ibid.

33. Goldsborough to Welles and Goldsborough to Lincoln, May 9, 1862, in Gideon Welles, *Report of the Secretary of the Navy in Relation to Armored Vessels* (Washington, D.C., 1864), 21–23.

34. Goldsborough to Welles, May 8, 12, 1862, in Welles, *Report of the Secretary in Relation to Armored Vessels,* 21–24.

35. Ibid.; Goldsborough to his wife, May 12, 1862, Goldsborough papers.

36. Goldsborough to Rodgers, May 7, 1862, ORN.

37. Keeler to his wife, May 12, 1862, Keeler Letterbook.

38. Ibid., May 15, 1862. *Monitor* log, May 11, 1862.

39. Jefferson Davis, quoted in *Civil War Naval Chronology,* part 2, *1862* (Washington, D.C., 1962), 38; Goldsborough to Welles, May 18, 1862, in Welles, *Report of the Secretary in Relation to Armored Vessels,* 24; John M. Coski, *Capital Navy: The Men, Ships, and Operations of the James River Squadron* (Campbell, Calif., 1996), chap. 2. The *Virginia*'s sailors did not actually engage the *Monitor.*

40. Keeler to his wife, May 15, 1862, Keeler Letterbook; Jeffers to Goldsborough, May 22, 1862, in Welles, *Report of the Secretary in Relation to Armored Vessels,* 27–29 (quotation on 28). *Monitor* log, May 15–16, 1862.

41. Keeler to his wife, May 15, 1862, Keeler Letterbook.

42. Ransford E. Van Gieson, quoted in Kurt Hackemer, "The Other Union Ironclad: The USS Galena and the Critical Summer of 1862," *Civil War History* 40, no. 3 (1994): 237–38.

43. Rodgers to Goldsborough and Newman to Rodgers, May 16, 1862, in Welles, *Report of the Secretary in Relation to Armored Vessels,* 25–26; Goldsborough to Welles, May 15, 16, 1862, in U.S. War Department, *The War of the Rebellion: A Compilation of the Official Records of the Union and Confederate Armies,* 128 vols. (Washington, D.C., 1885–1902); Hackemer, "The Other Union Ironclad," 234–37.

44. Keeler to his wife, May 16, 1862, Keeler Letterbook.

45. Ibid., May 15, 1862.

46. Ibid., May 12, 1862; Samuel Dana Greene, "In the Monitor Turret," in Johnson and Buel, *Battles and Leaders,* 729.

47. Keeler to his wife, May 22, 1862, Keeler Letterbook; Geer to his wife, May 20, Geer letters. The event is also described in the official reports. Smith to Goldsborough, May 19, 1862, ORN. *Monitor* log, May 19.

48. Keeler to his wife, June 18, 1862, Keeler Letterbook.

49. Patrick Malone writes about European colonists encountering Indian war-making in New England, but he notes that the term "skulking" was also used in Viet Nam. *The Skulking Way of War* (Baltimore, 1991); Geer to his wife, May 15, 1862, Geer letters. *Monitor* log, May 20, 1862.

50. Newton to Ericsson, June 20, 1862, Ericsson papers, NYHS, emphasis in the original.

51. Geer to his wife, May 20, 24, 29, June 9, 13, July 1, 5, 18, 20, 1862, Geer letters. *Monitor* log, June 12, 14, 1862. See also the temperature charts in the *Monitor* log, June 27, 1862.

52. Keeler to his wife, June 25, 1862, Keeler Letterbook.

53. Ibid., June 14, 1862.

54. Ibid., June 22, 1862. Geer mentions this fire as well in his letter of June 23, 1862, Geer letters.

55. Keeler to his wife, June 3, 1862, Keeler Letterbook. *Monitor* log, June 3, 1862.

56. Keeler to his wife, June 14, 1862, Keeler Letterbook. For the official report regarding capture of the *Teaser,* see Frank Moore, ed., *The Rebellion Record: A Diary of American Events* (New York, 1862), 5:548–49; Geer to his wife, June 18, 1862, Geer letters.

57. Alan Trachtenberg, *Reading American Photographs: Images as History: Matthew Brady to Walker Evans* (New York, 1989), 83.

58. Keeler to his wife, June 23, 1862, Keeler Letterbook.

59. Ibid., June 30, 1862.

60. Ibid., July 2, 1862.

61. Jeffers to Goldsborough, May 22, 1862, in Welles, *Report of the Secretary in Relation to Armored Vessels,* 27–29 (quotation on 28).

62. Ibid., 28–29.

63. Ericsson to Fox, May 28, 29, 1862, Ericsson papers, LC.

64. Keeler to his wife, June 14, 1862, Keeler Letterbook.

65. Geer to his wife, Aug. 13, 1862, Geer letters.

66. Keeler to his wife, Sept. 22, 1862, Keeler Letterbook.

67. Ibid., Aug. 16, 1862.

68. Geer to his wife, Aug. 20, 1862, Geer letters.

69. Robert MacBride, *Civil War Ironclads: The Dawn of Naval Armor* (New York, 1962), 91.

70. Geer to his wife, Sept. 1, 3, 7, 1862, Geer letters.

71. Welles, *Report of the Secretary in Relation to Armored Vessels,* 54–55; Roberts, *USS New Ironsides.*

72. Keeler to his wife, July 30, 1862, Keeler Letterbook.

73. Ibid., Oct. 6, 1862.

74. Hiram E. Sickles to his wife, Nov. 6, 1862, copy in private collection of William N. Still, Jr.

Chapter 7. Utilitarians View the *Monitor*'s Fight, 1862–1865

1. Keeler to his wife, Nov. 11, 1862, Keeler Letterbook. George Geer was responsible for the new blower engine on the berth deck. Geer to his wife, Nov. 11, 1862, Geer letters.

2. Keeler to his wife, Nov. 17, 1862, Keeler Letterbook.

3. Ibid., Dec. 28, 1864.

4. Ibid., Jan. 4, 6, 1863.

5. Frank Butts, *My First Cruise at Sea and the Loss of the Ironclad "Monitor"* (Providence, R.I., 1878).

6. Keeler to his wife, Jan. 6, 1863, Keeler Letterbook.

7. "The First Cruise of the 'Monitor' *Passaic,*" *Harpers New Monthly Magazine,* Oct. 1863, 577–95 (quotation on 578). (The name of the magazine had changed by the time of publication.)

8. Drayton to Welles, Jan. 1, 1863, Ericsson papers, NYHS.

9. Ibid., Jan. 2, 1862; Bright to Drayton, Jan. 2, 3, 1862, in Gideon Welles, *Report of the Secretary of the Navy in Relation to Armored Vessels* (Washington, D.C., 1864), 32–35. "First Cruise of the 'Monitor' *Passaic.*"

10. Drayton to Welles, Dec. 26, 1862, in Welles, *Report of the Secretary in Relation to Armored Vessels,* 31–32.

11. Ibid., Jan. 2, 1862, 32-35 (quotation on 33).

12. Gideon Welles, entry for Jan. 3, 1863, *The Diary of Gideon Welles* (Boston, 1911).

13. Welles to Ericsson, Jan. 6, 1862, Ericsson papers, NYHS.

14. Ericsson to Rear Adm. Smith, Jan. 1, 1863, Ericsson papers, NYHS. Ericsson to Fox, Dec. 26, 30, 1862, Fox papers.

15. Ericsson to Welles, Jan. 10, 1863, Ericsson papers, LC.

16. For Rodgers's description of the gale, see Rodgers to his wife, Jan. 22, 1863; for commendations, see Fox to Rodgers, Jan. 23, 1864; Paulding to Fox, Jan. 24, 1863, all in John Rodgers Papers, Library of Congress, Washington, D.C. Stimers to Rodgers, Jan. 24, 1863; Rodgers to Welles, Jan. 26, 1863; Rodgers to Ericsson, Jan. 24, 1863, all in Welles, *Report of the Secretary in Relation to Armored Vessels,* 42-45; Robert Erwin Johnson, *Rear Admiral John Rodgers 1812-1882* (Annapolis, Md., 1967), 232, 257.

17. Welles, *Report of the Secretary in Relation to Armored Vessels,* 55–183; "Ironclad Reports," John Dahlgren Papers, Library of Congress, Washington, D.C., box 18; Proceedings of the Court-Martial are reprinted in Frank M. Bennett, *The Steam Navy of the United States* (Pittsburgh, 1896), chap. 23.

18. On the *Weehawken* vs. the *Atlanta,* see Rodgers to DuPont, June 17, 1864, in Edward K. Rawson and Robert Wood, eds., *Official Records of the Union and Confederate Navies in the War of the Rebellion* (Washington, D.C., 1898): Welles, *Report of the Secretary in Relation to Armored Vessels,* 198–215; Johnson, *Rear Admiral John Rodgers;* Robert MacBride, *Civil War Ironclads: The Dawn of Naval Armor* (New York, 1962). On the *Weehawken's* loss, see Simpson to Dahlgren and Dahlgren to Welles, Dec. 6, 1863, in Welles, *Report of the Secretary in Relation to Armored Vessels,* 297–98. For the attack on Charleston, see Robert Schneller, Jr., *Quest for Glory: A Biography of Rear Admiral John A. Dahlgren* (Annapolis, Md., 1996); Donald Canney, *The Old Steam Navy,* vol. 2, *The Ironclads, 1842-1885* (Annapolis, Md., 1993).

19. Donald Nevius Bigelow, *William Conant Church & the Army and Navy Journal* (New York, 1952), 97, 153–54.

20. "A Word on the Other Side," *Army and Navy Journal,* Aug. 29, 1863.

21. "Iron-Clads and Wooden Vessels," *Army and Navy Journal,* Sept. 3, 1863.

22. Ironclad captains (Drayton, John Rodgers, George Rodgers, Ammen, Fairfax, Downes) to Navy Department, Apr. 18, 24, 1863, Rodgers papers.

23. Jeffers to Goldsborough, May 22, 1862, in Welles, *Report of the Secretary in Relation to Armored Vessels,* 27–29; Richard Albion, *Makers of Naval Policy 1798–1947* (Annapolis, Md., 1980), 106–7; Bernard Brodie, *Sea Power in the Machine Age* (Princeton, N.J., 1941), 208–9.

24. Dahlgren to Welles, Jan. 28, 1864, in Welles, *Report of the Secretary in Relation to Armored Vessels,* 587; after Hampton Roads, Dahlgren wrote the department, "For want of information it is not possible for us to form a trustworthy opinion, from the incidents of combat, of the relative powers, offensive and defensive, of the two vessels." "To the Navy Department by Captain Dahlgren, being one of the Documents from the Navy Department Accompanying the Annual Message of the President to Congress, December 1, 1862," Dahlgren papers; Shneller, *Quest for Glory,* 295–97.

25. Goldsborough to Welles, Feb. 26, 1864, in Welles, *Report of the Secretary in Relation to Armored Vessels,* 577.

26. Brodie, *Sea Power,* 207.

27. A list of Ericsson's articles, 1862-70, compiled by William Conant Church, can be found in Ericsson papers, LC.

28. Quoted in William Conant Church, *The Life of John Ericsson* (New York, 1890), 1:282.

29. Jeffers to Goldsborough, May 22, 1862, in Welles, *Report of the Secretary in Relation to Armored Vessels,* 27–29 (quotation on 28).

30. Ericsson to Fox, May 28, 29, 1862, Ericsson papers, LC; Ericsson, "The Building of the Monitor," in Robert Underwood Johnson and Clarence Clough Buel, eds., *Battles and Leaders of the Civil War,* vol. 1, *From Sumter to Shiloh* (New York, 1887), 735–36. Ericsson here was responding to Greene, who had made an observation similar to Jeffers's about the angle of fire around the pilothouse.

31. Ericsson to Welles, Feb. 8, 1863, in Welles, *Report of the Secretary in Relation to Armored Vessels,* 601.

32. Ericsson to Welles, June 15, 1863, in Welles, *Report of the Secretary in Relation to Armored Vessels,* 603.

33. Ericsson to Fox, Dec. 17, 1862, Fox papers, box 17; Ericsson to Welles, Jan. 8, 10, 1864, Ericsson papers, LC.

34. Ericsson to Welles, Feb. 24, 1863, quoted in Church, *Life of Ericsson* 2:24. For an account of the episode of the "light draught monitors," see Edward William Sloan III, *Benjamin Franklin Isherwood: Naval Engineer: The Years as Engineer in Chief, 1861–1869* (Annapolis, Md., 1965), chap. 3; Julia Stimers Dubrow, ed., *The "Monitor" and Alban B. Stimers* (Orlando, Fla., 1936); Bennett, *Steam Navy of the United States,* 484–93.

35. "The *Monitor* Question," *Army and Navy Journal,* July 23, 1864, 793. Ericsson responded to this article in the journal on Aug. 8, 1864, claiming the opposite, that he was pointing to practical evidence and his opponents resorted only to theory.

36. Ericsson to Welles, Feb. 8, 1863, in Welles, *Report of the Secretary in Relation to Armored Vessels,* 600.

37. Quoted in Church, *Life of Ericsson,* 1:282. Steam engineers, as it turned out, were no great proponents of monitors either, because the cramped spaces pushed their berths, always a matter of contention with line officers, into undesirable corners of the vessel, Monte A. Calvert, *The Mechanical Engineer in America: Professional Cultures in Conflict* (Baltimore, 1967), 248.

38. Ericsson to Welles, n.d., quoted in Church, *Life of Ericsson* 2:11. Emphasis added.

39. Ericsson to Stimers, Mar. 13, 1862, in Stimers Dubrow, *The "Monitor" and Alban B. Stimers.*

40. Ericsson to Lincoln, Aug. 2, 1862, quoted in Church, *Life of Ericsson,* 2:34.

41. Alex Roland, *Underwater Warfare in the Age of Sail* (Bloomington, 1978), 97, 111.

42. Rodgers to his wife, Mar. 13, 1865 quoted in Johnson, *Rear Admiral John Rodgers,* 276.

43. William McNeill, *The Pursuit of Power: Technology, Armed Force, and Society from A.D. 1000 to the Present* (Chicago, 1982), 270–74.

Chapter 8. Melville and the Mechanic's War

1. Both Ericsson and Hawthorne anticipated what I. F. Clarke called the "epoch of extrapolation," in which authors from Jules Verne to H. G. Wells predicted the course of military technology with imaginary nightmares of future warfare. I. F. Clarke, *Voices*

Prophesying War, 2d ed. (Oxford, 1992), 47; Thomas C. Leonard, *Above the Battle: War-Making in America from Appomattox to Versailles* (New York, 1978), chaps. 5–6.

2. Newton Arvin, *Herman Melville* (New York, 1950), 267.

3. Merton M. Sealts, Jr., *Melville's Reading* (Columbia, S.C., 1988).

4. Hennig Cohen, introduction to Herman Melville, *The Battle Pieces of Herman Melville,* ed. Hennig Cohen (New York, 1963), 16.

5. The correspondence between Melville's and Hawthorne's writings on the ironclads has been noted in Leo B. Levy, "Hawthorne, Melville, and the *Monitor,*" *American Literature* 37 (Mar. 1965): 33. See also Fredrick P. Kroeger, "Longfellow, Melville, and Hawthorne: The Passage into the Iron Age," *Illinois Quarterly* 33 (Dec. 1970): 37.

6. Edward Stessel argues that Melville's musing on the passing of the age of wooden ships expressed his own anxiety of aging and obsolescence. "Naval Warfare and Herman Melville's War against Failure," *Essays in Arts and Sciences* 10 (May 1981): 59–77; Joyce Sparer Adler, *War in Melville's Imagination* (New York, 1981), 143.

7. Herman Melville, "The *Temeraire,*" in Melville, *Battle Pieces.*

8. Melville, "In the Turret," in Melville, *Battle Pieces.*

9. Stanton Garner, *The Civil War World of Herman Melville* (Lawrence, Kans., 1993), 135.

10. On Melville in Hampton Roads, see Garner, *The Civil War World of Melville,* 303.

11. Guert Gansevoort to John Ericsson, Jan. 30, 1862, Ericsson papers, NYHS.

12. Garner, *The Civil War World of Melville,* 102, 132, 166–70.

13. Herman Melville, "Hawthorne and His Mosses," *Literary World,* Aug. 17, 24, 1850, 418.

14. Hershel Parker, *Herman Melville: A Biography,* vol. 1, *1819–1851* (Baltimore, 1996), 882-83.

15. Herman Melville, "The Bell Tower" and "The Paradise of Bachelors and the Tartarus of Maids," in Herman Melville, *Great Short Works of Herman Melville* (New York, 1969), 234, 215–16.

16. Herman Melville, *Moby-Dick; or, The Whale* (New York, 1982), 408.

17. Garner, *The Civil War World of Melville,* 101; Parker, *Herman Melville,* 260, 289; William N. Still, Jr., *Ironclad Captains: The Commanding Officers of the USS "Monitor"* (Washington, D.C., 1988), 37.

18. Melville, *White Jacket: Or the World in a Man-of-War* (Evanston, Ill.; Chicago, 1970), 374–75. Subsequent citations of this work appear in the text as page numbers in parentheses.

19. Parker, *Herman Melville,* 286.

20. White Jacket's love for the tops anticipates Ishmael's metaphysical reverie "The Mast Head" in *Moby-Dick:* "There you stand, a hundred feet above the silent decks, striding along the deep, as if the masts were gigantic stilts, while beneath you and between your legs, as it were, swim the hugest monsters of the sea, even as ships once sailed between the boots of the famous Colossus of Rhodes" (223).

21. Herman Melville, "A Utilitarian View of the *Monitor*'s Fight," in *Battle Pieces,* 69–71.

22. Ericsson to Stimers, Mar. 13, 1862, in Julia Stimers Dubrow, ed., *The "Monitor" and Alban B. Stimers* (Orlando, Fla., 1936).

23. William Conant Church, *The Life of John Ericsson* (New York, 1890), 2:150.

24. Stessel, "Herman Melville's War against Failure," 66. Stessel points out that Melville repeated his sentiments on the *Monitor* in the later poem, "Bridegroom Dick."

Conclusion: Mechanical Faces of Battle

1. Gerald Linderman, *Embattled Courage: The Experience of Combat in the American Civil War* (New York, 1987), 271–74 (quotation on 271).

2. Richard Albion, *Makers of Naval Policy 1798–1947* (Annapolis, Md., 1980), 199–202 (quotation on 199); Harold Sprout and Margaret Sprout, *The Rise of American Naval Power 1776–1918* (Princeton, N.J., 1946), 165; Benjamin Franklin Cooling, *Gray Steel and Blue Water Navy: The Formative Years of America's Military-Industrial Complex 1881–1917* (Hamden, Conn., 1979), chap. 1. On Isherwood's fall, see Edward William Sloan III, *Benjamin Franklin Isherwood: Naval Engineer: The Years as Engineer in Chief, 1861–1869* (Annapolis, Md., 1965), chap. 11; Peter Karstens, *The Naval Aristocracy: The Golden Age of Annapolis and the Emergence of Modern American Navalism* (New York, 1972), 66–67.

3. Robert Stanley McCordock, *The Yankee Cheesebox* (Philadelphia, 1938), 391.

4. Linderman, *Embattled Courage,* 284–89.

5. Ibid., 397–405.

6. *Merrimack* and *Monitor* Panorama Company, "A Comprehensive Sketch of the *Merrimack* and *Monitor* Naval Battle," New York, 1886, Allen collection.

7. Roy F. Nichols, "Introduction to the New Edition," Robert Underwood Johnson and Clarence Clough Buel, eds., *Battles and Leaders of the Civil War,* vol. 1 (1887; reprint, New York, 1956), iii–vii.

8. Samuel Dana Greene, "In the *Monitor* Turret," in Johnson and Buel, *Battles and Leaders.*

9. Worden to Welles, Jan. 5, 1868, NARA, HA, "Engagement with Enemy War Vessels, 1775–1862," box 174, folder "Accounts of Battle of Hampton Roads."

10. Jones to Fox, 1874, Fox papers; Southern Historical Society Papers, 1893, 2-3, Allen collection.

11. Ericsson to Fox, Nov. 24, 1874, NARA, HA, "Engagement with Enemy War Vessels, 1775–1862," box 174, folder "Accounts of Battle of Hampton Roads."

12. Welles to Fox, Sept. 25, 1875, Fox papers, box 17.

13. Stimers to Fox, Oct. 18, 1875; Greene to Fox, Nov. 15, 1875, both in Fox papers, box 17.

14. Butts to Pierce, Sept. 2, 1885; Durst to Pierce, May 28, 1885; White to Pierce, Sept. 6, 1886; Stodder to Pierce, July 28, 1886; Taylor to Pierce, Feb. 17, 1892, all in Pierce papers. Charles MacCord, Ericsson's draftsman, also reports the inventor's disappointment at the crew's performance in "Ericsson and his Monitor," *North American Review* 149, no. 4 (Oct. 1889): 469.

15. Keeler to Pierce, May 27, 1885, Feb. 9, 1886; Anna Keeler to Pierce, Dec. 4, 1884, May 23, 1886, all in Pierce papers; Robert W. Daly, ed., *Aboard the USS Florida: 1863–65: The Letters of Paymaster William Frederick Keeler, U.S. Navy, to His Wife, Anna* (Annapolis, Md., 1968).

16. Greene, "In the *Monitor* Turret," 729; John Ericsson, "The Building of the *Monitor*," in Robert Underwood Johnson and Clarence Clough Buel, eds., *Battles and Leaders of the Civil War,* vol. 1, *From Sumter to Shiloh* (New York, 1887), 737. Gideon Welles, "Healthiness of Ironclads," in *Report of the Secretary of the Navy* (Washington, D.C., 1865), xxi–xxii. Welles's annual report summarizes the surgeon general's report, although the

study itself has not been found in an extensive search of the archives. The reason for the lower sick reports could be that the monitors were never far from shore and hence had more fresh food, or that the iron structures were more sanitary than wood.

17. Ericsson, "Building of the *Monitor*" (quotation on 742), and Samuel Dana Greene, "In the *Monitor* Turret," both in Johnson and Buel, *Battles and Leaders*; *Army and Navy Journal*, Feb. 28, 1885; Louis Stodder to Pierce, Dec. 20, 1885, Pierce papers; Ericsson to Fox, Mar. 4, 1874, Fox papers.

18. Butts to Pierce, Sept. 2, 1885; Stodder to Pierce, Oct. 5, Dec. 20, 1885, all in Pierce papers (emphasis original).

19. Ericsson to Fox, Mar. 4, 1874, Fox papers.

20. Geer to his wife, Aug. 20, 1862, Geer letters; Stimers to his wife, Mar. 8, 1862, in Julia Stimers Dubrow, ed., *The "Monitor" and Alban B. Stimers* (Orlando, Fla., 1936).

21. John Ericsson, *Contributions to the Centennial Exhibition* (1876; reprint, Stockholm, 1976).

22. *New York Times*, Aug. 23, 1890; Ruth White, *Yankee from Sweden: The Dream and the Reality in the Days of John Ericsson* (New York, 1960), 259–66; William Conant Church, *The Life of John Ericsson* (New York, 1890), 2:324–25.

23. Gordon P. Watts, Jr., *Investigating the Remains of the U.S.S.* Monitor: *A Final Report on 1979 Site Testing in the Monitor National Marine Sanctuary* (Washington, D.C., 1982); Gordon P. Watts, Jr., "Exploring the *Monitor*: 1979 Archaeological Investigation of the Civil War Ironclad," in J. Lee Cox, Jr., and Michael A. Jehle, eds., *Ironclad Intruder: U.S.S.* Monitor (Philadelphia, 1988), 44–54; Gordon P. Watts, Jr., "USS *Monitor*," in James P. Delgado, ed., *Encyclopedia of Underwater Archaeology* (New Haven, Conn., 1997); Gordon P. Watts, Jr., "The Location and Identification of the Ironclad USS *Monitor*," *International Journal of Nautical Archaeology* 4, no. 2 (1975): 301–29; John Broadwater, *Current Diving Research at the* Monitor *National Marine Sanctuary: Utilizing Technology to Save a Historic Warship* (Washington, D.C., 1998).

24. *Monitor* National Marine Sanctuary, *Charting a New Course for the "Monitor"* (Washington, D.C., 1997).

25. Daniel Pedersen, "Saving a Sunken Treasure," *Newsweek*, Apr. 20, 1998, 58; William J. Broad, "Saving the Ironclad Ship that Revolutionized Warfare at Sea," *New York Times*, Dec. 2, 1997.

26. Richard Gould, *Recovering the Past* (Albuquerque, 1990), 193–94.

27. Ernest W. Peterkin, "The Importance of Understanding the Construction of the USS *Monitor*," in William B. Cogar, ed., *Naval History: The Seventh Symposium of the Naval Academy* (Wilmington, Del., 1988).

28. *The Monitor, Its Meaning and Future: Papers from a National Conference* [Raleigh, N.C., Apr. 2–4, 1978] (Washington, D.C., 1978).

29. Linderman, *Embattled Courage*, 134.

30. Daniel Aaron, *The Unwritten War: American Writers and the Civil War* (New York, 1973).

31. Paul Fussel, *The Great War and Modern Memory* (New York, 1975), 24.

32. Ibid., 71.

33. Ibid., 169.

34. Ibid., 115; Eric J. Leed, *No Man's Land: Combat and Identity in World War I* (Cambridge, 1979), chap. 4.

35. The case of the *Monitor* thus supports Daniel Pick's critique of Fussel that he overly

sentimentalizes the pre-1914 world as innocent of modernity's frights. Daniel Pick, *War Machine: The Rationalization of Slaughter in the Modern Age* (New Haven, Conn., 1993), 200–204.

36. Richard Haigh, *Life in a Tank* (Boston, 1918), 1, 5; J. Burnett-Stuart, "The Progress of Mechanizations," *Army Quarterly* 16 (Apr. 1928): 50; G. Murray Wilson, ed., *Fighting Tanks: An Account of the Royal Tank Corps in Action, 1916–1919* (London, 1929), 158, all quoted in Barton C. Hacker, "Imaginations in Thrall: The Social Psychology of Military Mechanization, 1919–1939," *Parameters* 12, no. 1 (1982): 51.

37. J. F. C. Fuller, "The Development of Sea Warfare on Land and Its Influence on Future Naval Operations," *Journal of the Royal United Service Institution* 65 (May 1920): 293; R Hilton, "Fire Power or Armour," *Journal of the Royal United Service Institution* 73 (Feb. 1928): 66, both quoted in Hacker, "Imaginations in Thrall," 52–53.

38. Peter Fritzsche, *A Nation of Fliers: German Aviation and the Popular Imagination* (Cambridge, Mass., 1992), 86; John H. Morrow, Jr., *The Great War in the Air: Military Aviation from 1909 to 1921* (Washington, D.C., 1993), xv, 365–66; Michael Sherry, *The Rise of American Air Power: The Creation of Armageddon* (New Haven, Conn., 1987), 20; Leed, *No Man's Land,* 104. Lee Kennett, *The First Air War, 1914–1918* (New York, 1991), argues that the attitude of men in the trenches toward airmen was more ambivalent than simple admiration (156–57).

39. Fritzsche, *A Nation of Fliers,* 99–100; Sherry, *The Rise of American Air Power,* 39. Kennett, *The First Air War,* chap. 9.

40. Richard Hallion, *The Rise of the Fighter Aircraft: 1914–1918* (Annapolis, Md., 1984), 72; Sherry, *The Rise of American Air Power,* 205. Kennett, *The First Air War,* presents statistics showing that the reality of training accidents, though significant, was more complex than commonly believed (128–29).

41. Fritzche, *A Nation of Fliers,* 49.

42. Ibid., 58.

43. Ibid., emphasis added.

44. Herbert C. Fyfe, *Submarine Warfare: Past, Present, and Future* (London, 1902), 14. Richard Compton-Hall, *Submarines and the War at Sea, 1914–18* (London, 1991), chap. 2, "Living in a Can." For a personal account of pre–World War I submarines, see Charles A. Lockwood, *Down to the Sea in Subs* (New York, 1967).

45. Fyfe, *Submarine Warfare,* 3; Panel on Psychology and Physiology, Committee on Undersea Warfare, *A Survey Report on Human Factors in Undersea Warfare* (Washington, D.C., 1949), 417.

46. Robert Gardiner, Roger Chesnau, and Eugene M. Kolensik, eds., *Conway's All the World's Fighting Ships 1860–1905* (New York, 1980), 119–23.

47. Keeler to his wife, Dec. 9, 1862, Keeler Letterbook.

Bibliographical Essay

This essay briefly surveys those books that proved most relevant to this study and are easily available to the reader who may wish to probe further. In my attempt to bring the notion of experience to the study of military technology, I draw both on recent work by historians of technology and on John Keegan's *The Face of Battle: A Study of Agincourt, Waterloo, and the Somme* (London, 1976), the path-breaking work on soldiers' experience. It has a fine introductory essay on the historiography of battle. For the American Civil War, Gerald Linderman's *Embattled Courage: The Experience of Combat in the American Civil War* (New York, 1987) looks at soldiers' experience of the land war and how the incessant battle of 1864–65 affected the values with which they entered the conflict. James McPherson, one of today's best-known Civil War historians, addresses similar topics in *For Cause and Comrades: Why Men Fought in the Civil War* (New York, 1997), explicitly modeling the book on Keegan, although coming to different conclusions from Linderman's. Both draw on the classic social histories of Civil War soldiers by Bell Irvin Wiley, *The Life of Billy Yank: The Common Soldier of the Union* (Indianapolis, 1952), and *The Life of Johnny Reb: The Common Soldier of the Confederacy* (Indianapolis, 1943).

On general approaches to studying technology, Merritt Roe Smith and Leo Marx's edited volume, *Does Technology Drive History? Essays on Technological Determinism* (Cambridge, Mass., 1994), contributes to an understanding of technological determinism and of the problems of "autonomous" technological development, problems that arise when one speaks of the *Monitor* in revolutionary terms. Donald MacKenzie's *Inventing Accuracy: A Historical Sociology of Nuclear Missile Guidance* (Cambridge, Mass., 1993) employs similar ideas in his precise analysis of a modern weapons system. MacKenzie demonstrates that there is no "natural trajectory" of development inherent in a technology and shows that disputes arise over even basic measures of performance. On the cultural history of early American technology, Leo Marx's classic *The Machine in the Garden: Technology and the Pastoral Ideal in America* (London, 1964) examines the relationship between Americans' ideologies of nature and their views of machinery; Marx's discussion of Hawthorne's and Melville's responses to machinery and industrialization is the background for my analysis of their writing on the *Monitor*. In a similar vein is Rosalind Williams's investigation into the technology and imagery of the underground, from subways to tunnels to military fortifications. Williams argues in *Notes on the Underground: An Essay on Technology, Society, and the Imagination* (Cambridge, Mass., 1990) that images of the underground represented artificial environments in a future dominated by technology, an argument that can be applied as well to the enclosed world of the *Monitor*. Judith McGaw's book on papermaking in the Berkshires, *Most Wonderful Machine: Mechanization and Social Change in Berkshire Paper Making, 1801–1885* (Princeton, N.J., 1987), gives an excellent account of a high-technology industry in the region in which Hawthorne and Melville lived. Also see Margaret Creighton's social history of American whaling, *Rites and Passages: The Experience of American Whaling, 1830–1870* (Cambridge, 1995). Whaling had much in common with naval experience. And see Margaret Creighton and Lisa Norling's edited volume, *Iron Men, Wooden Women: Gender and Seafaring in the Atlantic World 1700–1920* (Baltimore, 1996), on the gender dimensions of seafaring.

Military histories that delve deeply into society and technology are few, but the literature is growing. Merritt Roe Smith's book *Harper's Ferry Armory and the New Technology: The Challenge of Change* (Ithaca, N.Y., 1977) remains a classic account, revealing the struggles over mechanization and craft skills in antebellum small arms production. Ken Alder's more recent work on French artillery engineers, *Engineering the Revolution: Arms and Enlightenment in France, 1763–1815* (Princeton, N.J., 1997), argues that roots of uniformity and instrumental rationality in machine production lay in prerevolutionary France. In Merritt Roe Smith's edited volume, *Military Enterprise and Technological Change: Perspectives on the American Experience* (Cambridge, Mass., 1985), historians write about military technology not simply as a parade of weapons and battles, but as products of human organizations whose values and internal conflicts affect development (the book also includes a valuable bibliographic essay by Alex Roland on technology and war). William McNeill's sweeping study *The Pursuit of Power: Technology, Armed Force, and Society from A.D. 1000 to the Present* (Chicago, 1982) does much to put the nuclear arms races into the historical perspective of "command technology," wherein governments learned to direct the growth of machinery toward their own ends.

Other historians trace the connections between military technology and mainstream culture in ways relevant to the *Monitor*'s story. I. F. Clarke's *Voices Prophesying War,* 2d ed. (Oxford, 1992), shows how fantastic visions of future technological wars have influenced policymakers since the mid–nineteenth century. Peter Fritzsche's *A Nation of Flyers: German Aviation and the Popular Imagination* (Cambridge, Mass., 1992) adds much to our understanding of the public imagery of fighter aces, as does Lee Kennett's *The First Air War, 1914–1918* (New York, 1991). Michael Sherry undertakes a similar analysis for strategic bombing in *The Rise of American Air Power: The Creation of Armageddon* (New Haven, Conn., 1987). Also see Thomas C. Leonard, *Above the Battle: War-Making in America from Appomattox to Versailles* (New York, 1978), chapters 5–6, and John H. Morrow, Jr., *The Great War in the Air: Military Aviation from 1909 to 1921* (Washington, D.C., 1993). Cultural histories of modern war that bring together representation (in the form of letters, music, and literature) and soldiers' experience tend to concentrate on World War I. The best is Paul Fusell's *The Great War and Modern Memory* (New York, 1975), although other fine volumes by Modris Eksteins (*Rites of Spring: The Great War and the Birth of the Modern Age* [Boston, 1989]), Daniel Pick (*War Machine: The Rationalization of Slaughter in the Modern Age* [New Haven, Conn., 1993]), Eric J. Leed (*No Man's Land: Combat and Identity in World War I* [Cambridge, 1979]), and Samuel Hynes (*The Soldier's Tale: Bearing

Witness to Modern War [New York, 1997]) address similar questions. In all of these accounts, as with the *Monitor,* the lines between "the military," "technology," and "culture" break down upon close analysis.

Newton Arvin has written biographies of both Hawthorne and Melville that have stood the test of time (*Nathaniel Hawthorne* [Boston, 1929]; *Herman Melville* [New York, 1950]). James Mellow's *Nathaniel Hawthorne in His Times* (Baltimore, 1988) makes a fine read, as well. Hershel Parker's long-awaited Melville biography, *Herman Melville: A Biography,* vol. I, *1819–1851* (Baltimore, 1996), though only the first part, serves as an encyclopedic reference on the author's life. Stanton Garner's *The Civil War World of Herman Melville* (Lawrence, Kans., 1993) is most relevant to Melville's writing and experience during the 1860s. It details the writer's extensive family network and its connections to the war. Melville's *White Jacket: Or the World in a Man-of-War* (Evanston, Ill.; Chicago, 1970) remains the most readable and detailed account of life in the antebellum navy.

On the early generation of technical officers in the navy, see the biography of Isherwood by Edward William Sloan III, *Benjamin Franklin Isherwood: Naval Engineer: The Years as Engineer in Chief, 1861–1869* (Annapolis, Md., 1965), the biography of Dahlgren by Robert Schneller, Jr., *Quest for Glory: A Biography of Rear Admiral John A. Dahlgren* (Annapolis, Md., 1996), and James C. Bradford's edited volume, *Captains of the Old Steam Navy: Makers of the American Naval Tradition, 1840–1880* (Annapolis, Md., 1986). *From Sail to Steam: Recollections of Naval Life* (New York, 1906), by Alfred Thayer Mahan, the nineteenth century's most influential naval strategist, details his life as a midshipman before and during the Civil War and treats technological change from the point of view of a junior officer. Frank M. Bennett's history of nineteenth-century naval engineering, *The Steam Navy of the United States* (1896; reprint, Westport, Conn., 1972), contains useful detail on the engineering corps and its struggles to attain legitimacy within the navy. David K. Brown's *Before the Ironclad* (London, 1990), written by an experienced marine engineer, traces the antebellum debates over naval warfare. Similarly, Andrew Lambert's *Battleships in Transition: The Creation of the Steam Battlefleet, 1815–1860* (London, 1984) shows that the wooden-hulled, steam-powered warship (such as the *Merrimack*) lasted much longer than historians usually recognize. These works taken together, especially with the volume *Steam, Steel, and Shellfire: The Steam Warship 1815–1905* (London, 1992), edited by Robert Gardiner (which contains essays by Brown, Lambert, Still, and others), make the standard revolutionary account of ironclad warships more complex than a simple series of revolutionary inventions. Alex Roland discusses another technological trajectory rife with exaggerated visions and failure, but with profound military and personal implications, in *Underwater Warfare in the Age of Sail* (Bloomington, Ind., 1978), including material on Robert Fulton, a strange military mind in the mold of John Ericsson. For Ericsson himself, few book-length studies exist. William Conan Church's lengthy, admiring *Life of John Ericsson* (New York, 1890) needs to be taken with several grains of salt, though it is well written and worth reading (see chapter 2 for further comments). Ruth White's more recent, popular biography, *Yankee from Sweden: The Dream and the Reality in the Days of John Ericsson* (New York, 1960), is only slightly less celebratory; a comprehensive, critical look at Ericsson awaits a modern scholar.

For many years the naval aspects of the Civil War received scant attention from historians. That lack has been remedied in recent decades, but naval subjects still have not generated the literature that army operations and land warfare have. Bern Anderson's *By Sea and by River: The Naval History of the Civil War* (New York, 1962) has for years served as a standard account (it includes a historiographical essay in the introduction). The history of confederate armorclads by William N. Still, Jr., *Iron Afloat: The Story of the Confederate Armorclads* (Nashville, Tenn., 1971), shows that the Southern vessels were as innovative as those built in the North. Dennis J. Ringle has recently written a social history of the Civil War navy, *Life in Mr. Lincoln's Navy* (Annapolis, Md., 1998), which concentrates on daily life. It is mostly compiled from personal accounts. Robert Bruce's *Lincoln and the Tools of War* (1954; reprint, Chicago, 1989) details Lincoln's curious involvement in weapons design and development and his friendship with John Dahlgren. Gideon Welles's personal diary (*Diary of Gideon Welles,* vol. 1 [Boston, 1911]) is a window into the higher levels of administration, as is John Niven's biography of Welles, *Gideon Welles: Lincoln's Secretary of the Navy* (New York, 1973). Robert Erwin Johnson's biography of John Rodgers, *Rear Admiral John Rodgers 1812–1882* (Annapolis, Md., 1967), adds valuable material about the *Monitor's* summer up the James River. It also recounts Rodgers's own command of a monitor. John M. Coski's recent study of the James River Squadron, *Capital Navy: The Men, Ships, and Operations of the James River Squadron* (Campbell, Calif., 1996), is valuable here as well. William H. Roberts's USS *New Ironsides in the Civil War* (Annapolis, Md., 1999) sheds light on the *Monitor's* politics and on its most successful competitor. The voluminous *Official Records of the Union and Confederate Navies in the War of the Rebellion* (Washington, D.C., 1898), edited by Edward K. Rawson and Robert Wood, is an easily available standard documentary reference on the subject and surprisingly interesting reading for such a vast collection.

On the *Monitor* and the *Virginia,* the literature is vast, although largely repetitive of the standard accounts. Recent examples of standard accounts include William C. Davis, *Duel between the First Ironclads* (Baton Rouge, 1975); A. A. Hoehling, *Thunder at Hampton Roads* (New York, 1993); Edward M. Miller, USS *"Monitor:" The Ship That Launched a Modern Navy* (Annapolis, Md., 1978); James Tertius deKay, *"Monitor"* (New York, 1997). Practically every five years, a new book has appeared, usually repeating and solidifying the canonical account. Nonetheless, some works stand above the others. James Phinney Baxter's *Introduction of the Ironclad Warship* (Cambridge, Mass., 1933) conclusively debunked the revolutionary mythology (although scholarly revision did little to stem celebration of the "schoolboy's" *Monitor*). Bernard Brodie's *Sea Power in the Machine Age* (Princeton, N.J., 1941) builds on Baxter's work, adding a more strategic perspective. Brodie's work takes on added historical interest since he became one of the leading strategists of nuclear war, applying his insights into nineteenth-century military technology to the twentieth century (Baxter, too, brought the *Monitor's* lessons to bear on this century, writing the official history of American research and development during World War II, *Scientists against Time*). Robert Daly's *How the Merrimack Won: The Strategic Story of CSS Virginia* (New York, 1957) provides a detailed look at the strategic circumstances surrounding Hampton Roads and makes a strong argument that the *Virginia* had significant effects on the war effort.

Other useful works include Edward M. Miller's *U.S.S. Monitor: The Ship That Launched a Modern Navy* (Annapolis, Md., 1978), which contains a bibliography of the numerous books, articles, speeches, memoirs, and analyses published about the *Monitor.* Donald Canney, in *The Old Steam Navy,* vol. 2, *The Ironclads, 1842–1885* (Annapolis, Md., 1993), provides a technical discussion of the monitor class. Robert Stanley McCordock's *The Yankee Cheesebox* (Philadelphia, 1938), though not the strongest account of the ship itself, has wide coverage of newspaper responses in the North, the South, and Europe (see especially chapter 17). *Battles and Leaders of the Civil War*, vol. 1 (1887; reprint, New York, 1956), edited by Robert Underwood Johnson and Clarence Clough Buel, is a popular series of firsthand accounts originally published in *Century* magazine. Hampton Roads appears in the first volume, *From Sumter to Shiloh*, although its articles must be understood in the context of their time (see the conclusion to this volume). William Keeler's letters were published in 1964 and 1968, edited and with informative annotations by Robert Daly (*Aboard the U.S.S. Monitor, 1862: The Letters of Acting Paymaster William Frederick Keeler, U.S. Navy, to His Wife Anna* [Annapolis, Md., 1964]; *Aboard the USS Florida: 1863–65: The Letters of Paymaster William Frederick Keeler, U.S. Navy, to His Wife, Anna* [Annapolis, Md., 1968]). Also relevant is the memoir of Thomas Oliver Selfridge, Jr., who was aboard the *Cumberland* during its defeat by the *Virginia* and who briefly commanded the *Monitor* (*What Finer Tradition: The Memories of Thomas O. Selfridge, Jr., Rear Admiral, U.S.N.* [Columbia, S.C., 1987]). Thomas Alva Hunter's life aboard the monitor *Nahant* reveals that the discomforts of those vessels continued well after the demise of the original *Monitor*. See his memoir *A Year on a "Monitor" and the Destruction of Fort Sumter*, edited by Craig L. Symonds (Columbia, S.C., 1987).

The discovery of the *Monitor* wreck prompted a number of conferences and publications that include interesting essays and commentaries. Especially valuable is their consideration of the role of the *Monitor* in popular culture and their incorporation of the wreck and its discovery into the standard accounts. See, for example, *The Monitor, Its Meaning and Future: Papers from a National Conference Raleigh, North Carolina, April 2–4, 1978* (Washington, D.C., 1978); J. Lee Cox, Jr., and Michael A. Jehle, eds., *Ironclad Intruder: U.S.S.* Monitor (Philadelphia, 1988). Gordon P. Watts, Jr.'s reports on the excavations (*Investigating the Remains of the U.S.S.* Monitor: *a Final Report on 1979 Site Testing in the Monitor National Marine Sanctuary* [Washington, D.C., 1982] and "The Location and Identification of the Ironclad USS *Monitor*," *International Journal of Nautical Archaeology* 4, no. 2 [1975]: 301–29) describe the wreck site and the artifacts recovered, and the most recent report from the National Marine Sanctuary that oversees the wreck site (John Broadwater, *Current Diving Research at the* Monitor *National Marine Sanctuary: Utilizing Technology to Save a Historic Warship* [Washington, D.C., 1998]) describes several alternative plans for raising and preserving the ship. As part of the sanctuary program, the U.S. government sponsored a number of publications on the *Monitor*, the most useful being the detailed examinations of its builders and captains by William N. Still, Jr. (Monitor *Builders: A Historical Study of the Principal Firms and Individuals Involved in the Construction of USS* Monitor [Washington, D.C., 1988]; *Ironclad Captains: The Commanding Officers of the USS "Monitor"* [Washington, D.C., 1988]).

Index

LIBRARY OF CONGRESS CATALOGING-IN-PUBLICATION DATA

Mindell, David A.

   War, technology, and experience aboard the USS Monitor / David A. Mindell.

      p.   cm.

   Includes bibliographical references and index.

   ISBN 0-8018-6249-3 (alk. paper : alk. paper)—ISBN 0-8018-6250-7 (pbk. : alk. paper)

   1. Monitor (Ironclad)   2. United States. Navy—History—Civil War, 1861–1865.

3. United States—History—Civil War, 1861–1865—Naval operations.   I. Title.

VA65.M65 M55 2000

973.7'52—dc21   99-038344